INTRODUCING
THE PLANETS AND THEIR MOONS

Other Titles in this Series:

Introducing Astronomy (2014)
Introducing Geology – A Guide to the World of Rocks (Second Edition 2010)
Introducing Geomorphology (2012)
Introducing Meteorology ~ A Guide to the Weather (2012)
Introducing Mineralogy (2014)
Introducing Oceanography (2012)
Introducing Palaeontology – A Guide to Ancient Life (2010)
Introducing Sedimentology (2014)
Introducing Tectonics, Rock Structures and Mountain Belts (2012)
Introducing Volcanology ~ A Guide to Hot Rocks (2011)

For further details of these and other Dunedin
Earth and Environmental Sciences titles see
www.dunedinacademicpress.co.uk

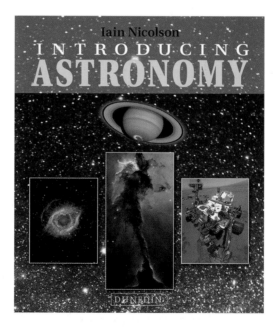

INTRODUCING

THE PLANETS AND THEIR MOONS

Peter Cattermole

DUNEDIN

EDINBURGH ◆ LONDON

This book is dedicated to Roger Jones,
with whom I published several books and learned
so much about the business of writing.

Published by Dunedin Academic Press Ltd

Head Office: Hudson House,
8 Albany Street, Edinburgh EH1 3QB

London Office: The Towers,
352 Cromwell Tower, Barbican, London EC2Y 8NB

www.dunedinacademicpress.co.uk

ISBNs
9781780460291 (Paperback)
9781780465135 (ePub)
9781780465142 (Kindle)

British Library Cataloguing in Publication data
A catalogue record for this book is available from the British Library

Design and pre-press production by Makar Publishing Production, Edinburgh
Printed and bound in Poland by Hussar Books

Contents

Preface

The opportunity to write this book about planets and moons could not be missed. When I was a raw undergraduate I had for many years enjoyed a fascination for the Moon and planets, engendered by one of the most enthusiastic of scientists, Sir Patrick Moore. Sadly he is no longer with us, but he, like me, was able to enjoy the amazing explosion of knowledge that followed the first landing on the Moon in 1969. This amazing feat was followed by the incredible journeys of Voyager into the most distant reaches of the Solar System, exploration of both Mars and Venus by a number of highly successful probes; journeys to Mercury, trips around the Sun, and currently robotic vehicles that are analysing the rocks, minerals and dust on the surface of Mars. As recently as March 2014, it was announced that a 450 km-diameter dwarf planet, named 2012 VP113, interestingly of pinkish hue, had been discovered well beyond the orbit of Pluto, adding further weight to the argument that a large number of irregular icy bodies lie out in the distant reaches of our planetary system.

As a result of all this activity, our understanding of what makes up our local bit of the Galaxy has improved tremendously. We now know that there are far more moons than we had suspected; that more planets have ring systems; and we have an appreciation of what the surfaces, interiors and atmospheres of many of these worlds are like. Having at one time been a part of NASA's research effort, I know what an amazing achievement this has been and how a large number of dedicated scientists and technicians are responsible for the understanding that we now have.

This book is an introduction to current ideas about the planets and their moons. It is not an exhaustive treatment; this would require a far longer book. Rather, I have attempted to give a flavour of what the larger members of our planetary system are like, and if, in so doing, it inspires my readers to delve deeper into this fascinating subject, then I will have achieved my ambition.

Peter Cattermole

Note: I have tried to minimise the use of technical terms, but those that are necessary are explained in the Glossary and are highlighted in **bold** type on first appearance.

List of tables and illustrations

1 Introduction

Until very recently, many scientists held the view that the **Solar System** was unique. In part this was due to the fact that carbon-based life had evolved on Earth, and in part because astronomers had been unable to detect any other planets in our **Universe**. Today, however, planetary systems have been discovered elsewhere in our own **Galaxy** and must, by any kind of logic, exist in others. There is thus little reason to suppose that some form of life has not developed there also, and that *Homo sapiens* and the other forms of life that flourish here, are not unique.

The Solar System comprises a central **star** – the Sun – and a large number of much smaller, denser, bodies that include the eight **planets**: Mercury, Venus, Mars, Earth, Jupiter, Saturn, Uranus and Neptune, together with their moons, **dwarf planets** and large numbers of **meteoroids**, **asteroids** and **comets** (Figure 1.1). Most of the smaller bodies orbit the Sun in the same plane – known as the **ecliptic** – and the entire system rotates and moves through Space. In fact the Sun and its attendant family take roughly 200 million years to rotate around the centre of our Galaxy, known as the Milky Way.

Stars usually are composed of hydrogen, deuterium, tritium, helium, and lithium and have a mass that is sufficient to sustain stable

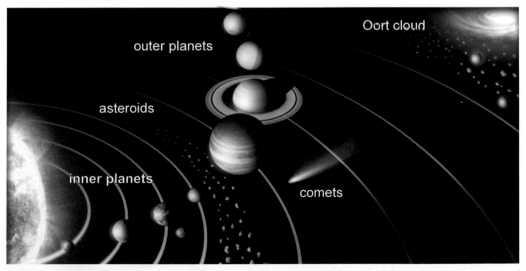

Figure 1.1 The Solar System consists of the Sun, eight planets, several dwarf planets, asteroids, comets and meteoroids. Adapted after NASA.

fusion reactions. Because of these nuclear reactions, they emit massive amounts of **electromagnetic radiation** at a wide range of wavelengths. Planets, on the other hand, are usually relatively cool and stable, and much smaller. They may be small, rocky bodies, such as the **terrestrial planets**, dwarf planets, and asteroids, or much larger bodies, known as giant planets, composed predominantly of gases and ices.

Planets, being relatively non-massive, are gravitationally bound to more massive stars, which is the situation in our own Solar System. During the early stages of its evolution, many of the planets captured smaller bodies that now orbit around them; these are their moons. Amongst them are Earth's Moon, the Galilean satellites of Jupiter and many others. Such bodies have a wide range of size and composition.

Each of the giant planets has a **ring system**, the one orbiting Saturn being particularly large and spectacular. Such ring systems are composed of myriad small ice and rock particles. Completely separate smaller bodies, unrelated to the major planets, include asteroids, meteoroids and comets. Asteroids are predominantly rocky bodies that orbit the Sun between the orbits of Mars and Jupiter, and may represent particles from a disrupted planet (Figure 1.2). Comets are made largely of ice and are to be found orbiting the Sun beyond the orbit of Neptune in what is termed the **Kuiper Belt**, and also well beyond all of the major planets in a diffuse spherical zone known as the **Oort Cloud**.

Meteoroids are rocky fragments that range in size from sand to small boulders and represent debris left over from the formation of the Solar System. The unaltered silicate

Figure 1.2 Galileo mission image of the asteroids Ida (left) and Gaspra (right). The former measures 29 km long, the latter 17 km. Courtesy NASA/JPL.

group known as **chondrites** contains some of the oldest known fragments of Solar System material, with ages of 4.56 Ga. Many scientists consider these to represent the primordial matter from which the rocky planets were made. When such fragments pass through the terrestrial atmosphere they become heated and may impact the Earth as **meteorites**.

Once there were considered to be nine planets, the outermost, Pluto, being discovered as recently as 1930. However, astronomers were not sure about Pluto's mass until the discovery in 1978 of an attendant companion, named Charon. Calculations on the orbital behaviour of the two enabled astronomers to establish that Pluto had a diameter of 2400 km, which was puzzling, as it was far too small to cause certain orbital perturbations that had been observed.

However, powerful new ground- and space-based observations have completely changed our understanding of the outer Solar System. Instead of being the only planet in its region, Pluto and its moon are now known to be examples of a collection of objects that orbit the Sun within the Kuiper Belt, a region that extends from the orbit of Neptune out to 55 **astronomical units**. Astronomers estimate that there are at least 70 000 icy objects in this region similar in composition to Pluto, and many of these are more than 100 km across. As a consequence, Pluto/Charon was demoted to the class of dwarf planet. There are four other such substantial dwarfs, namely Makemake, Eris, Haumea and Ceres.

Scientists had supposed that extrasolar planets existed for many years, but it was not until 1988 that the first published discovery was made by Bruce Campbell, G. A. H. Walker, and Stephenson Yang. They were cautious about claiming detection of a planetary body orbiting the star Gamma Cephei and many scientists refuted its existence. Two years later further observations supported their claim, but it was not until 2003, when improved techniques were developed, that the planet's existence was confirmed.

In April 1992, two radio astronomers announced the discovery of two planets

Figure 1.3 This composite image shows the first planet outside of our solar system (right) found orbiting a brown dwarf, dubbed 2M1207 (centre). Image obtained by the ESO Paranal Observatory.

orbiting the **pulsar** PSR 1257+12. Subsequently, in 1995, the first confirmed detection of a planet orbiting a main-**sequence** star was made, when a giant planet was found in a 4-day orbit around the nearby star 51 Pegasi. As of January 2013, a total of 859 such planets, in 676 planetary systems, including 128 multiple planetary systems, have been identified (Figure 1.3). The **Kepler mission** alone has detected over 18 000 additional candidates, including 262 potentially habitable ones.

In February 2014, NASA's Kepler mission announced the discovery of 715 new planets. These newly verified worlds orbit 305 stars, revealing multiple-planet systems much like our own solar system. Nearly 95% of these planets are smaller than Neptune.

2   The origin of the Solar System

Cosmologists believe that the Sun and its attendant family of planets originated in a large **molecular cloud** within our local galaxy. This cloud of interstellar gas and dust became disturbed, possibly by the shock wave from a nearby **supernova** explosion, and collapsed under its own gravity. As the cloud collapsed it heated up and became compressed towards its centre, so much so that the dust vaporized. The initial collapse probably took less than 100 000 years.

The solar nebula

Because of the competing forces associated with gravity, gas pressure, and rotation, the contracting **nebula** began to flatten into a spinning pancake shape with a bulge at the centre. As it collapsed further, instabilities in the collapsing, rotating cloud caused local regions to contract gravitationally. These local regions of condensation eventually became the Sun and the planets, as well as their moons and the smaller fragments that are a part of the Solar System (Figure 2.1).

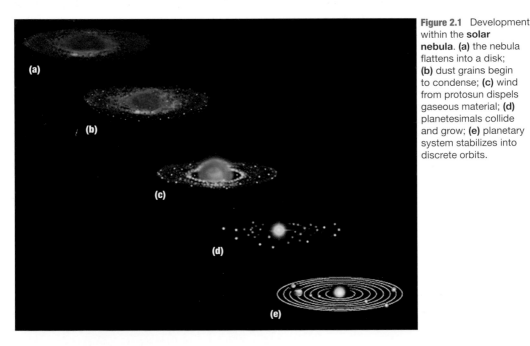

Figure 2.1 Development within the **solar nebula**. **(a)** the nebula flattens into a disk; **(b)** dust grains begin to condense; **(c)** wind from protosun dispels gaseous material; **(d)** planetesimals collide and grow; **(e)** planetary system stabilizes into discrete orbits.

As time passed, the spinning disk contracted more, whereupon the central object, at that stage a rather unstable T Tauri star, ignited and evolved into the Sun. Meanwhile, in the outer part of the nebula, gravity caused matter to condense around density perturbations and dust particles, and it began separating into rings. This was the stage at which runaway accretion caused successively larger fragments of dust and debris to become clumped together as proto-planets. Earth formed in this manner about 4.54 Gya and the process of formation was largely completed within 10–20 million years.

Because the nebula was warmer close to the proto-Sun, heavier elements could remain there, but the more volatile ones were swept outward towards its edge. The high-energy flow of particles from the early T-Tauri star – the **solar wind** – was responsible for sweeping the lighter elements towards the outer part of the disk. Perhaps less than a million years after the system was born, and when the proto-Sun was still surrounded by the dense disk of gas and dust, the gas giant planets Jupiter and Saturn formed. However, once the gas had been expelled from the system, the likelihood of more giant planets forming diminished. The terrestrial planets would have taken longer to form: growth proceeding through multiple violent collisions for millions of years after most of the original gas had left the system.

Modern research suggests that between 650 and 700 million years after the formation of the Solar System, Jupiter and Saturn entered orbits where the period of Saturn was exactly twice that of Jupiter. This **orbital resonance** caused strong interactions within the disk because of the continuously repeating gravitational effects from these two massive planets. As a result, the outer planets were shifted into their current orbits. Because the overall gravitational field of the system was in a state of flux, there was a phase when most of the smaller bodies that had not yet been captured into planets were thrown around. As these bodies shifted orbits, some scientists have hypothesized that there was a period of intensive impact about 3.85 Gya, called the **Late Heavy Bombardment**. However, this topic is a real 'hot potato' and currently is being vigorously debated.

One of the issues revolves around the fact that life started on Earth around 3.85 Gya, which would coincide with this very hazardous episode. It appears unlikely that life could evolve while Earth was being so vigorously impacted. So, in one camp are those that believe the solar system experienced a cataclysm of large impacts about 3.8 Gya, while in the other are those who prefer to believe that impacts were spread more evenly over time, from approximately 4.3 to 3.8 Gya.

Condensation from the solar disk

The early separation of dust, ice and gas eventually led to the development of an inner group of relatively dense rocky worlds – Earth, Mercury, Venus and Mars – and an outer group of much less dense, but very large, gas and ice giants: Jupiter, Saturn, Uranus and Neptune. Currently the orbits of these bodies are stable, but they did not become so until the two-to-one resonance between Jupiter and Saturn was reached. It was at this stage that Uranus and Neptune were sent into their current positions.

As the planetary disk cooled, the earliest dust grains to form would have been those of **refractory compounds** followed by compounds with successively lower condensation temperatures. Eventually, the dust is predicted to have consisted of olivine, feldspar and oxidized iron minerals.

The most primitive samples we have of solar disk material are believed to be the carbonaceous chondrite meteorites. These ancient fragments preserve a record of early events in our Solar System (at least the inner part) and yield radiometric ages of 4.56 Ga. The **C1-chondrite** group of meteorites – a class of stony meteorite – are the most abundant and consist of three kinds of material: (a) refractory inclusions formed of calcium and aluminium minerals; (b) **chondrules** – spherical particles composed largely of olivine and pyroxene; and (c) a matrix rich in hydrated silicates, clay minerals and organic compounds (Figure 2.2).

Their chemistry is not the same as that of the solar **photosphere**, but their non-gaseous elemental abundances very closely resemble it. The presence of high- and low-temperature phases indicates that prior to accretion, the primordial solar nebula consisted of a mixture of refractory and non-refractory phases. Asteroid-sized bodies must have been forming not just in the asteroid belt but everywhere in the solar system. They would have begun aggregating into larger bodies in a process that eventually produced the rocky inner planets, a process that was remarkably rapid.

The Moon probably formed by an impact of a Mars-sized body with the growing Earth. The oldest Moon rocks that have been dated are about 4.44 Ga, but there is evidence that the Moon actually formed within 30 Ma of the refractory inclusions. Similarly, the oldest meteoritic material from Mars is about 4.5 Ga old, but there is evidence that Mars itself formed about 13 Ma after the refractory inclusions. Thus, within as little as 30 Ma of the

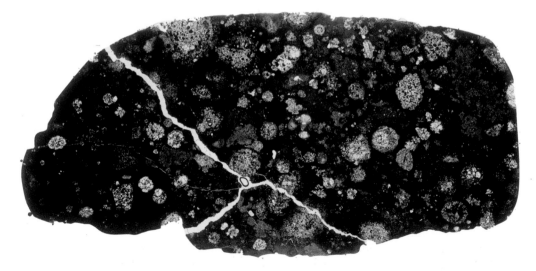

Figure 2.2 Section through the Allende meteorite, a typical carbonaceous chondrite. Courtesy NASA/JPL.

appearance of the first solids, the aggregation process that started with tiny particles had produced the rocky inner planets.

The planets, therefore, did not all develop in the same way; some formed close to the Sun and swept up the denser solid pieces, while others settled much further out and condensed largely from ice and gas, forming the giant outer planets. So, very early on, there were two distinct planet families. Furthermore, there was a widely different potential to store energy: the greater the mass of a body, the greater the potential it had to store it. With a wide range of planetary dimensions, e.g. Jupiter r = 72 000 km; Mercury r = 2440 km; Earth r = 6371 km, it is not surprising that the developmental history of the various planets was very different.

Planetary accretion and heating

Once the various proto-planets had begun to form, their growing masses exerted a gravitational attraction on the myriad smaller pieces of matter that were whirling about. This gave rise to the 'early bombardment', a kind of disk sweeping-up process. It gave rise to **accretion** and it played a vital role in the evolution of the Sun's family. This phase of heavy bombardment left its mark on all of the rocky worlds and on all of the satellites of the planets that had Moons. It is particularly well seen on our own Moon, whose heavily cratered surface bears the imprint of this violent collisional stage; but the Moon is not alone and nearly every rocky body is pockmarked with craters. Recent pictures of Mercury show this early impact record superbly (Figure 2.3).

Planetary surfaces are affected by two types of process: internal (e.g. volcanism) and external (e.g. impact). Impact and accretion

Figure 2.3 MESSENGER image of the limb of Mercury, showing the light-coloured Rembrandt basin. PIA17370. Courtesy NASA/JPL.

of incoming particles was responsible for the gradual growth of the planets, and in addition contributed huge amounts of heat as each impactor carried considerable kinetic energy that was transferred to the impacted body as heat (Figure 2.4).

Over millions of years this is believed to have raised Earth's temperature by as much as 4000°C. Furthermore, as the protoplanets grew in size, their interiors became gradually warmer due to the natural **adiabatic gradient**. The present temperature and pressure gradient is shown in Figure 2.5.

Figure 2.4 Incoming bodies.

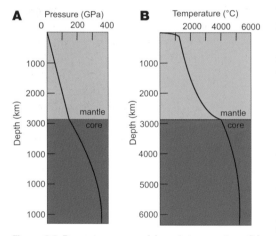

Figure 2.5 Present pressure **(a)** and temperature **(b)** gradients inside the Earth.

Planet size and position with respect to the central Sun was critical. Earth and Venus are significantly larger than both Mars and Mercury and so retained more accretional heat than their smaller neighbours, and also lost it much more slowly. Venus, being much nearer to the Sun than the Earth, lost most of its volatiles, retaining the heavy gas carbon dioxide, which accumulated to form a dense, choking atmosphere. Earth was sufficiently far away from the Sun to retain significant volatiles, which in due course enabled our world to develop an atmosphere and hydrosphere that provided conditions suitable for life. On the other hand, Mars, being much further from the Sun's heat, effectively froze its atmosphere into the sub-crust, and furthermore, because of its small size, lost its internal heat far more rapidly.

Jupiter, Saturn, Uranus, and Neptune formed beyond the frost line, the point between the orbits of Mars and Jupiter where the disk temperature was cool enough for volatile icy compounds to remain solid. The ices that formed the Giant Planets were more abundant than the metals and silicates that formed the terrestrial planets, allowing them to grow massive enough to capture hydrogen and helium, the lightest and most abundant elements. Today, the four gas giants comprise nearly 99% of all the mass orbiting the Sun.

It has been suggested that both comets and carbonaceous chondrites formed beyond the orbit of Jupiter, perhaps even at the edges of our Solar System, and then moved inward, eventually bringing their volatiles and organic material to Earth. This notion has recently been questioned.

The asteroid belt

Recent research indicates that the asteroid belt is found today in a dramatically different state than that immediately following its formation, and is estimated to have been depleted in total mass by a factor of at least a thousand. The asteroid belt also hosts a wide range of compositions, with the inner regions dominated

Figure 2.6 View of the asteroid Vesta's north pole, imaged by NASA's Dawn spacecraft. Courtesy NASA/JPL-Caltech/UCLA/MPS/DLR/IDA.

by S-type and other water-poor asteroids and the outer regions dominated by C-type and other primitive asteroids (Figure 2.6).

A recently-proposed model of early inner solar system evolution – colloquially known as the 'Grand Tack' – proposes that the gas-driven migration of Jupiter and Saturn bought them closer to the Sun, thereby truncating the disk of **planetesimals** in the terrestrial planet region, before migrating outwards toward their current locations. Thus, the orbital migration of Jupiter caused a very early depletion of the asteroid belt region, which was then repopulated from two distinct source regions, one inside the formation region of Jupiter and one between and beyond the giant planets. The scattered material then reformed the asteroid belt, producing a population of objects with the appropriate mass, orbits, and with overlapping distributions of material from each parent source region.

The most common bodies are the **C-type asteroids**, forming around 75% of those known. Such bodies are composed of silicates but are also rich in carbonaceous compounds, and in consequence have very low reflectivity. They tend to dominate the outer part of the asteroid belt beyond 2.7 AU. *S-types* are made from silicates but lack the dark carbon compounds and thus are more reflective. Most of them appear to be primitive and give ages of 4.65 Ga (Figure 2.7). They make up a smaller fraction of the asteroids than the C-types. Gaspra and Ida, explored by the Galileo spacecraft on its way to Jupiter, and Eros, recently orbited by the NEAR spacecraft, are examples of this type. The third group are the **M-type asteroids** which are composed of metals like iron and nickel. These latter are much rarer than the others and may represent core fragments of small differentiated objects.

The solar system has had a complex developmental history, but is now in a relatively stable configuration and will likely continue that way until the Sun reaches its old age. It is now time to briefly consider planetary orbits.

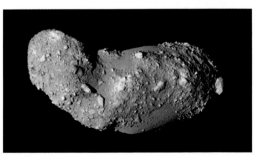

Figure 2.7 The small asteroid Itokawa. Courtesy NASA/JAXA.

3 Planets and their orbits

Planets and moons travel in elliptical orbits about a point known as the **barycentre,** a point at the centre of a system, weighted according to the distribution of mass within it. The Moon and the Earth rotate about a common point within the Earth but not near the centre; Jupiter and the Sun rotate about a common point just outside the solar surface (Figure 3.1). Over time the **eccentricity** of the various orbits varies, due to gravitational interactions between the various bodies. Generally, the orbits do not depart much from circularity.

All of the planets circle the Sun more or less in the same plane, which is termed the ecliptic and is defined by the plane of the Earth's orbit. The ecliptic is inclined by only 7° from the plane of the Sun's equator. Figures 3.2(a) and 3.2(b) show the relative sizes of the orbits of the eight planets plus the dwarf planet Pluto. They all orbit counter-clockwise, looking down from above the Sun's North Pole. All but Venus, Uranus and Pluto also rotate in that same sense. The Pluto/Charon system has an orbit that is significantly inclined to the ecliptic plane.

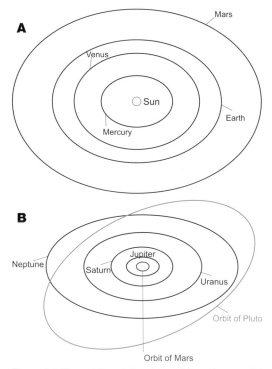

Figure 3.2 The orbits of the inner **(a)** and outer **(b)** planets. All of the major planets have orbits that lie in the plane of the ecliptic. However, that of dwarf planet Pluto is inclined to that plane.

Figure 3.1 The barycentre (B) of the Earth-Moon (E, M respectively) system lies within the Earth.

Mercury

Mercury's orbit is eccentric, and therefore the surface receives very different amounts of insolation when at different points in its path. Early astronomers believed it had a **synchronous rotation**, but in 1965 a non-synchronous period was confirmed. The true period is 58.6 days – two-thirds of the orbital period. The axial inclination is negligible, so that Mercury spins in an almost 'upright' position.

Since the rotation period (58.7 Earth days) is equal to two-thirds of the revolution period (88 Earth days) it follows that to an observer at a fixed point on Mercury the interval between one sunrise and the next will be 176 Earth days, or 2 Mercurian years (Figure 3.3). The spin–orbit relationship, together with the eccentricity of 0.20563, means that at alternate **perihelion** passages, the same hemisphere faces the Sun; likewise at **aphelion** passages. Because of this relationship, the central meridians of 0° and 180° always face the Sun at perihelion passage; in consequence at its closest approach the surface temperature of Mercury is at its most extreme (740K).

One of these 'hot spots' lies within the Caloris Basin. At the 90° and 270° meridian, maximum temperatures are about 200K lower. Of all the planets in the solar system, Mercury shows the greatest range in temperature.

Venus

Venus moves around the Sun in a practically circular orbit at a mean distance of 0.723 AU and has a period of 224.7 Earth days. It is the only planet to have a rotation period longer than its orbital period. It has a **retrograde motion**, and because of this the length of a solar day on Venus (116.75 Earth days) is significantly shorter than the **sidereal day**.

Venus may have formed from the solar nebula with a very different rotation period and obliquity, reaching its current state because of chaotic spin changes caused by planetary perturbations, impacts or tidal effects on its dense atmosphere. However, we have no information that informs us about this aspect of its behaviour.

Earth

Our own planet is the largest of the group of rocky planets, and has a rotation period of 23 h 56 m 4 s. Its orbit is almost circular, with a mean distance from the Sun of 149,597,870 km. The rotational axis is tilted to the orbital plane at an angle of 23°. However, all three of the orbital properties, i.e. axial tilt, obliquity and eccentricity vary. The variations in orbital characteristics were closely studied by Serbian mathematician Milutin Milankovitch back in the 1930s, as he felt that they could play an important role in Earth's changing climate. Collectively known as the Milankovitch Cycles, they affect the seasonality of solar radiation reaching the Earth's surface (Figure 3.4).

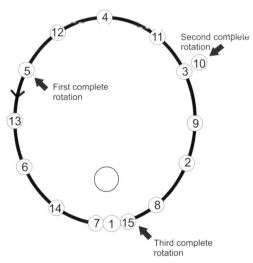

Figure 3.3 Mercury's rotation period is exactly two-thirds the length of the orbital revolution. As a result, for an observer on the surface, the interval between successive sunrises will be two Mercurian years.

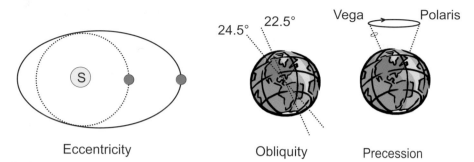

Eccentricity   Obliquity   Precession

Figure 3.4 The Milankovitch cycles. The precessional cycle is 23000 years long, the axial tilt 41000 years long, and the ellipticity changes over a 100000 year period.

The shape of the Earth's orbit around the Sun is constantly fluctuating, orbital shape ranging between more and less elliptical (0 to 5% ellipticity) in a cycle of about 100000 years. These oscillations, from more to less elliptical, are of prime importance as they alter the distance from the Earth to the Sun, thus changing the distance the Sun's short-wave radiation must travel to reach Earth, consequently reducing or increasing the amount of radiation received at the Earth's surface in different seasons.

Oscillations in the degree of Earth's axial tilt occur on a periodicity of 41000 years from 21.5 to 24.5°. Today the Earth's axial tilt is about 23.5°, which largely accounts for our seasons. One hypothesis for Earth's reaction to a smaller degree of axial tilt is that it would promote the growth of ice sheets. The third of the cycles is **precession**, which is the Earth's slow wobble as it spins on its axis. It changes from pointing at Polaris to pointing at the star Vega. This top-like wobble has a periodicity of 23000 years. Due to this, a climatically significant alteration must take place. When the axis is tilted towards Vega the positions of the northern hemisphere winter and summer solstices will coincide with the aphelion and perihelion, respectively. This means that the northern hemisphere will experience winter when the Earth is furthest from the Sun, and summer when the Earth is closest to the Sun. This coincidence will result in greater seasonal contrasts.

The Moon

The Moon's rotation is synchronous, which means that the same area of the Moon is turned towards the Earth all the time, although the eccentricity of the lunar orbit leads to **libration** zones which are brought alternately in and out of view. From Earth only 59% of the Moon's surface can be studied during the course of each year, while only 41% is permanently out of view. Tidal forces have been responsible for this relationship, which is not unusual; most other planetary satellites also have synchronous rotation with respect to their primaries. The barycentre of the Earth–Moon system lies 1707 km beneath the terrestrial surface, so that the statement that 'the Moon moves round the Earth' is not really misleading. Because of tidal effects, the Moon is receding from the Earth at a rate of 3.83 cm/year.

Mars

Mars has a special place in the history of physics and astronomy, as it was study of the movements of Mars during the sixteenth century that allowed the German mathematician, Johannes Kepler, to formulate his three **Laws of Planetary Motion**, the first of which was published in 1609. This derived from the fact that when he measured the distance of Mars from the Sun at different points in its orbit, it was apparent that the path followed was not a circle, but an ellipse.

Mars has a more eccentric orbit than Earth (eccentricity 0.093), being 207 million kilometres from the Sun at perihelion and 249 million kilometres at aphelion. The seasons on Mars are of the same general type as those of Earth, since the axial tilt is very similar and the Martian day (termed a *sol*) is not a great deal longer (1 sol = 1.029 days). Southern summer occurs near perihelion, therefore climates in the southern hemisphere of Mars show a wider range of temperature than those in the north. The effects are much greater than for Earth, partly because there are no seas on Mars and partly because of the greater eccentricity of the Martian orbit. At perihelion, Mars receives 44% more solar radiation than at aphelion.

Because Mars's southern hemisphere is inclined towards the Sun at perihelion, spring and summer seasons in that hemisphere are respectively 52 and 25 Earth days shorter than in autumn and winter. The greater eccentricity of Mars's orbit also has an effect on the maximum summer temperatures felt by the two hemispheres. At present, southern summers are shorter and warmer than those in the north; this is because the planet is 20% nearer to the Sun at perihelion than at aphelion, and it therefore receives 45% more incoming insolation (Figure 3.5).

Minor bodies (asteroids and dwarf planets)

Asteroid and dwarf planet orbits are of various definite types. Those within the *Main Belt* circle the Sun between the orbits of Mars and Jupiter. Then there are others, known as the *Inner Asteroids*, that orbit closer in, and include the class known as near-Earth asteroids (NEAs). Closer to the Sun are the *Aten* class bodies whose orbits have a mean distance from the Sun of less than 1 AU; some may cross the Earth's orbit (Figure 3.6). All of these bodies are small.

Apollo class bodies orbit at a mean distance from the Sun greater than 1 AU, but their orbits do cross that of the Earth. *Amor* class asteroids

Mars farther from Sun and moving slowly

Mars closer to Sun and moving faster

S. hemisphere winter long and cold
N. hemisphere summer long and cool

N. hemisphere winter brief and mild
S. hemisphere winter brief and hot

Figure 3.5 The seasons on Mars are more extreme than those of Earth, largely due to the planet's greater eccentricity.

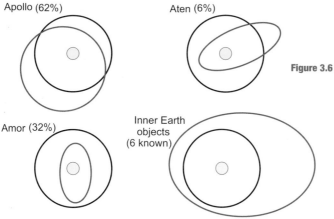

Apollo (62%)   Aten (6%)

Amor (32%)   Inner Earth objects (6 known)

Figure 3.6 Orbital relationships of asteroid groups.

circle in orbits that cross the path of Mars, but not that of the Earth. The group known as the *Trojans* have a more specific characteristic: each moves in the same orbit as a planet, though keeping on average either 60° ahead of the planet or 60° behind. Jupiter has hundreds of Trojans attendants, some of which are large by asteroid standards. Neptune has several, but none have yet been found associated with either Saturn or Uranus. Mars has several small Trojans, but Earth, Venus and Mercury apparently do not.

Centaur objects have orbits with perihelia greater than that of Jupiter but less than that of Neptune. The first to be found was 2060 Chiron, which moves mainly between the orbits of Saturn and Uranus; this has a diameter in excess of 100 km. Another Centaur, 5145 Pholus, has an orbit crossing the paths of Saturn, Uranus and Neptune.

Jupiter

Jupiter is the only planet that has a centre of mass with the Sun that lies outside the volume of the Sun, though by only 7% of the Sun's radius. The average distance between Jupiter and the Sun is 778 million kilometres (5.2 AU) and it completes an orbit every 11.86 years. This is two-fifths the orbital period of Saturn, meaning that there is a 5:2 orbital resonance between the two largest planets in the Solar System. Because of an eccentricity of 0.048, the distance from Jupiter and the Sun varies by 75 million kilometres between perihelion and aphelion. The elliptical orbit of Jupiter is inclined 1.31° compared to the Earth.

The axial tilt of Jupiter is relatively small: only 3.13°. As a result it does not experience significant seasonal changes. Jupiter's rotation is the fastest of all the Solar System's planets, completing an axial rotation in slightly less than ten hours; this creates an equatorial bulge easily seen through an Earth-based telescope. Thus Jupiter's equatorial diameter is 9,275 km longer than that measured through the poles. Because the planet is not a solid body, its upper atmosphere undergoes differential rotation, the rotation of the polar atmosphere being about 5 minutes longer than that of the equatorial.

Saturn

The average distance between Saturn and the Sun is over 1.4 billion kilometres (9 AU). With an average orbital speed of 9.69 km/s, it takes Saturn about 29½ years to complete one revolution. Its orbit is fairly regular, being a mere 0.06 from circular and inclined 2.48° relative to the orbital plane of the Earth. Because of an eccentricity of 0.056, the distance between Saturn and the Sun varies by approximately 202 million kilometres between perihelion and aphelion. A day on Saturn is less than half that of a day on Earth, and a year is about 30 times longer. That comes to more than 24,232 Saturnian days in the Saturnian year. More interesting than the details of Saturn's orbit are the objects that orbit around it: its moons and rings (Figure 3.7).

A precise value for the rotation period of the interior remains elusive. The latest estimate of Saturn's rotation based on a compilation of various measurements from the Cassini, Voyager and Pioneer probes was reported in September 2007 as 10 h, 32 m, 35 s.

Uranus

Uranus revolves around the Sun once every 84 Earth years, at an average distance of roughly 3 billion kilometres (about 20 AU). Consequently the intensity of sunlight on Uranus is about 1/400 that on Earth. Its orbital elements were first calculated in 1783 by Pierre-Simon Laplace; subsequently discrepancies began to appear between the predicted and observed orbits, and in 1841, John Couch Adams proposed that the differences might be due to the gravitational tug of an unseen planet. In 1845, Urbain Le Verrier began his own independent research into Uranus's orbit, and on 23 September 1846, Johann Gottfried Galle

Figure 3.7 The cloud belts and rings of Saturn, imaged by Voyager 2. Three of Saturn's icy moons are evident; they are in order of distance from the planet: Tethys, 1050 km (652 mi) in diameter; Dione, 1120 km (696 mi); and Rhea, 1530 km (951 mi). The shadow of Tethys appears on Saturn's southern hemisphere. PIA23887. Courtesy NASA/JPL.

located a new planet, later named Neptune, at nearly the position predicted by Le Verrier.

The rotational period of the interior of Uranus is 17 h, 14 m, and is retrograde. As on all the giant planets, its upper atmosphere experiences very strong winds in the direction of rotation. At some latitudes, visible features of the atmosphere move much faster, making a full rotation in as little as 14 hours.

The planet has an axial tilt of 97.77°, so its axis of rotation is approximately parallel with the plane of the Solar System. Uranus rotates rather like a tilted rolling ball, which gives it seasonal changes completely unlike those of the other major planets. Near the time of Uranian solstices, one pole faces the Sun continuously while the other pole faces away. For this reason only a narrow belt around the equator experiences a rapid day–night cycle, but with the Sun very low over the horizon. At the other side of Uranus's orbit the orientation of the poles towards the Sun is reversed. Each pole gets around 42 years of continuous sunlight, followed by 42 years of darkness. Near the time of the equinoxes, the Sun faces the equator of Uranus giving a period of day–night cycles similar to those seen on most of the other planets. One result of this axis orientation is that, on average during the year, the polar regions of Uranus receive a greater energy input from the Sun than the equatorial.

Neptune

The average distance between Neptune and the Sun is 4.5 billion kilometres (about 30.1 AU), and it completes an orbit on average every 164.79 years. In July 2011, Neptune completed its first full orbit since its discovery in 1846, although it did not appear at its exact discovery position in our sky because the Earth was in a different location in its orbital cycle. The elliptical orbit of Neptune is inclined 1.77° and has an eccentricity of 0.011, therefore the distance between Neptune and the Sun varies by 101 million kilometres between perihelion and aphelion. The axial tilt of Neptune is 28.32°, which is similar to the tilts of both Earth

and Mars; as a result, it experiences similar seasonal changes. The long orbital period means that the seasons last for 40 Earth years. Its day is roughly 16.11 hours long.

As with the other gas giants, its atmosphere undergoes differential rotation. The wide equatorial zone rotates with a period of about 18 hours, which is slower than the 16.1-hour rotation of the planet's magnetic field. By contrast, the reverse is true for the poles where the rotation period is 12 hours. This differential rotation is the most pronounced of any planet in the Solar System, and it results in strong latitudinal wind shear.

Neptune's orbit has a profound impact on the region directly beyond it, known as the Kuiper belt. This ring of small icy worlds, similar to the asteroid belt but far larger, extending from Neptune's orbit at 30 AU out to about 55 AU from the Sun, is dominated by Neptune's gravity. Over the age of the Solar System, certain regions of the Kuiper Belt became destabilized, creating gaps in its structure.

There do exist orbits within these empty regions where objects can survive for extremely long periods, perhaps as long as the age of the Solar System itself. The most heavily populated resonance in the Kuiper Belt, with over 200 known objects, is the 2:3 resonance. Objects in this resonance complete 2 orbits for every 3 of Neptune, and are known as *plutinos* because they are the largest of the known Kuiper Belt objects. Pluto is among them. Although Pluto crosses Neptune's orbit regularly, the 2:3 resonance ensures they can never collide.

4   Planetary differentiation

The planets, their moons and other bodies such as meteorites and asteroids were derived from the same nebula that spun out into the solar disk. This probably had about 0.15 solar masses, but was injected by material from a nearby exploding supernova captured by the Sun. Subsequently, many bodies evolved within it, with a wide range of chemical compositions. The process whereby this was achieved is known as **differentiation**, a fundamental geological process.

Once the original nebula had begun to clear due to the Sun's early T Tauri activity, (T~2000 K), the inner part of the cloud would have become depleted in H and He. This was the beginning of the process of differentiation. As cooling progressed, so condensation temperatures were reached and microscopic dust grains started to form. Such grains would have been in equilibrium with the surrounding gas, and the earliest minerals that formed would have been the more refractory ones. These were followed in a predictable sequence of increasing volatility.

The mineralogical sequence, assuming a temperature of 1700K, would have begun with compounds containing tungsten, osmium, zirconium and oxides of calcium, titanium and aluminium, followed by minerals similar to those now found in the carbonaceous chondrites, while at lower temperatures, iron and nickel would have condensed. As temperature fell to between 1400K and 1300K, Mg-silicates would have appeared, followed by sodium and potassium silicates, i.e. feldspars. By the time the nebular temperature had dropped to around 500K, the dust would have been composed of olivine, feldspar and oxidized iron minerals (Figure 4.1).

At temperatures still lower, carbon would have condensed; then water vapour would have reacted with some of the dust to produce hydrated minerals such as serpentine, tremolite and talc. At 200K, ice crystals would have appeared and in the outer reaches of the nebula, ammonia and methane ices. Some of

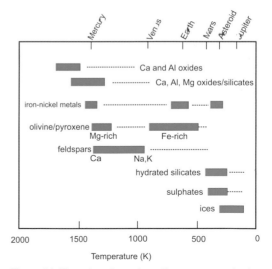

Figure 4.1 The mineral condensation sequence in the solar nebula.

the latter would have reacted with water to form hydrated ices and, because the heat of the Sun would have been able to sublimate ice that was unshielded in the inner Solar System, water ice became restricted to the outer parts of the system.

Chemistry of the Sun and Solar System

Each body condensed from the Solar Nebula at a different distance from the Sun; in consequence, the degree of solar heating each experienced, once in a stable orbit, varied. Furthermore, relative proximity to the Sun dictated whether or not particular elements could remain in the vicinity as they grew. Thus there was a primary differentiation of the elements, the more volatile elements being concentrated in the outer regions, and the denser elements nearer to the Sun. Subsequently individual planets fractionated their constituent elements into layered structures, generating planetary cores, mantles and crusts.

The aggregation of matter from which the Sun, planets and their moons evolved can be split into three components: gas, ice and rock. The gas was predominantly hydrogen (H) and helium (He); together they account for 98% of the initial mass of the nebula. Of the ices, water ice (H_2O) predominates, with gases such as ammonia (NH_3), methane (CH_4) and carbon dioxide (CO_2) accounting for the rest.

The chemical elements are classified according to their volatility but also are divided into **siderophile**, **chalcophile** or **lithophile** groups, depending upon their propensity for combining with metal, sulphide or silicate phases. Where scientists talk of 'solar system abundances', they are referring to the composition of all the material within the system, including the Sun. The present interstellar medium will be enriched in heavy elements compared with the matter from which the solar cloud developed, around 4.6 Gya.

Hydrogen is the most abundant chemical element in our universe, with helium coming second. These two are calculated to make up 74% and 24% respectively. The abundances of chemical elements in the solar atmosphere are shown in table 4.1.

The objects that once were considered to represent most closely the primordial nebular material are the C1 carbonaceous chondrites, a variety of the stony-iron meteorites (Figure 4.2). They are composed mainly of silicates, oxides and sulphides, and are characterized by the presence of the minerals olivine and serpentine. Several groups contain high percentages of water (3% to 22%), as well as organic

Table 4.1 Abundances of elements in the solar photosphere (after Unsöld, 1950).

Element	Atomic number	Abundance (atoms/10^4 atoms Si)
H	1	5.1×10^8
He	2	1×10^8
C	6	10 000
N	7	21 000
O	8	2.8×10^6
Na	11	1000
Mg	12	17 000
Al	13	1100
Si	14	10 000
S	16	4300
K	19	81
Ca	20	870
Sc	21	1.1
Ti	22	47
V	23	5.9
Cr	24	200
Mn	25	150
Fe	26	27 000
Co	27	55
Ni	28	470
Cu	29	8.7
Zn	30	31

Figure 4.2 The Orgeuil meteorite. This primitive object contains spherical chondrules, refractory inclusions and a fine-grained silicate matrix. Courtesy NASA/JPL.

compounds. The presence of volatiles implies that they have not undergone significant heating since they formed.

It is certainly true that we can use this group of objects as a guide to solar nebula composition, but recent research shows that fractionation of elements has modified the inner regions of the nebula, because the rocky planets are not of C1 composition. Evidently the inner worlds formed from material that had been depleted in elements volatile below 1700 K.

Comets also have been cited as being possibly representative of this elusive original composition. However, analysis of the core of Comet Halley showed that while it matched quite well solar photosphere abundances, two important elements, namely Si and Fe, diverged sharply. However, detection of crystalline Mg-rich olivine and orthopyroxene grains in the dust of the comet's inner **coma**, means that such silicate material must have already existed around certain types of stars that supply interstellar dust, because temperatures in the region of the Solar System from

whence comets derived are unlikely to have been sufficiently high for such grains to form.

Planetary heating

The early bombardment that began about 4.5 Gya gave rise to accretion of the Sun's family. It left its mark on all of the rocky worlds and on all the satellites of all the planets that had moons. It is particularly evident on Earth's Moon, whose heavily cratered surface bears the imprint of this violent planet-modifying stage (Figure 4.3).

As has been mentioned before, accretion was responsible for the growth of the planets and also implanted heat through transfer of kinetic energy. For example, a body of radius 10 metres, travelling towards Earth at a cosmic velocity of 10 km/sec, would generate energy

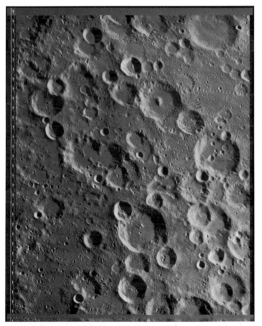

Figure 4.3 Lunar Orbiter 4 image of the cratered farside of the Moon. Courtesy NASA/JPL.

equivalent to 380 kilotonnes of TNT, or 19 Hiroshima bombs, or a scale 5 earthquake! Much of this heat would, of course, have been lost to space, but each time an impact occurred, particularly a sizeable one, some of the heat energy was added to the developing planets. Over millions of years this caused the Earth to heat up, possibly by as much as 6000 K. Because much of that heat would be stored in the outermost layers, it is likely that most planets developed **magma oceans**.

In addition, the increase in internal temperature brought about by the natural adiabatic gradient which, for the Earth, is at the rate of 2°C per 100 metres, would, for a body about three-quarters the size of the modern Earth, have raised the temperature at the planet's core by 900°C. As discussed in Chapter 2, planet size and position with respect to the central Sun were also critical, affecting the amount of volatiles retained, and whether an atmosphere and hydrosphere could develop.

Another event that generated immense amounts of heat via **gravitational potential energy** was the formation of planetary cores. At some point, probably within about 50 Ma of Earth's formation, the internal temperature was sufficiently high for metallic iron (Fe) to sink as molten droplets through the silicate mantle, towards Earth's centre. The Fe-rich droplets gradually settled to form the 2940 km-diameter core, made of iron (Fe) and nickel (Ni). In so doing they released a massive amount of energy that is believed to have raised the temperature of the Earth by a further 2000°C. A similar process would have operated within the other planets. Once the dense, hot and partially molten core formed, it became surrounded by the largely silicate mantle layer.

From the beginning, a mixture of light and heavy elements was incorporated into the rocky planets. Some of these elements were not only dense, but also radioactive; yet, despite being 'heavy', they did not sink down towards the Earth's core, as one might expect, because of their affinity for oxygen. Since oxygen became concentrated in the terrestrial crust, the heavy radioactive elements migrated upwards towards the surface, there to combine with lithospheric rocks.

Once these heavy radioactive elements (e.g. U, Pb, Sr, Nd) became locked into the minerals found in near-surface rocks they commenced to decay and an immense amount of heat was released. To this day, it is the long-lived **radionuclides** that provide a major source of internal heat. We believe that early in Earth's history, their concentration in the outer layers of the Earth may have meant that an ocean of molten rock, a magma ocean, encased our planet.

Differentiation into a layered structure

Dense metallic cores are assumed to have formed on all of the rocky planets at an early stage. This differentiation process, eventually producing a series of concentric shells of different composition, produced on Earth three principal layers: core, mantle and crust. It is assumed that this also occurred within the other members of the inner solar system and, while layered structures developed within the gas and ice giants, not all of the three layers are necessarily present.

Earth

Earth scientists have benefited from analysis of seismic waves and other high-quality geophysical and geochemical data to constrain

Figure 4.4
Harzwburgite nodule
from Olmani Hill,
Arusha, Tanzania.
A mantle nodule
brought up in an
explosive eruption.
Author photo.

the internal structure and composition of the Earth. But it should be remembered than no one has drilled a hole deeper than 12 km into the outermost crust (that is 0.2% of the way to the planet's centre). However, insights into the subsurface materials can sometimes be gained when xenoliths of deep-seated rocks are brought up in explosive eruptions, usually within regions of thinned continental crust (Figure 4.4).

The study of earthquake waves, whose behaviour is determined by the density and physical properties of the rocks through which they pass, has allowed geophysicists to establish the nature of Earth's deep interior (Figure 4.5). Beneath Earth's outer skin is the mantle layer, which is believed to be made from peridotite. The lower mantle

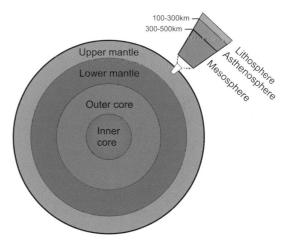

Figure 4.5 The internal structure of the Earth, showing a cut-out of the uppermost part of the mantle – the mesosphere, asthenosphere and lithosphere. The crust varies between 7 and 70 km in thickness.

contains 72.9% of the mantle-crust mass and is composed mainly of silicon, magnesium, and oxygen, together with iron, calcium, and aluminium. Above this lies the transition zone, or **mesosphere,** which is the source of the basaltic magmas that emerge at oceanic ridges. This layer is dense when cold, but is buoyant when hot because some of its contained minerals melt easily to form basalt, which can then rise through the upper layers as magma. The upper mantle, which often can be sampled in eroded mountain belts and in explosive volcanic deposits, is composed of the minerals olivine $(Mg,Fe)_2SiO_4$ and pyroxene $(Mg,Fe)_2Si_2O_6$. These and the other minerals it contains are refractory and crystallize at high temperatures. The part of the upper mantle called the **asthenosphere** may be partially molten.

The mantle stretches downwards all the way to the boundary of the outer liquid core, which is made largely from metallic iron and nickel, plus some sulphur. The outer core is constantly stirred by the Earth's rotation and this drives the planet's internal dynamo.

The melting of the rising mantle material is brought about by the release in pressure that it experiences when it approaches the surface. The magma that forms will contain the matter that has relatively lower melting points, which generates a melt of basaltic composition. Such melts form at about 1100°C. Basalt is the most widespread lava in the entire Solar System, forming the extensive plains on the Moon, Mercury, Venus and Mars.

Today 70% of the Earth is covered by ocean. The underlying oceanic crust is composed of basalt with a density of $3.1 \, gm/cm^3$, over which there is a thin veneer of sediment, together forming a layer on average 8 km thick. Because of its density, this crust generally sits lower than the adjacent continents. Continental crust, on the other hand, has a density of only $2.8 \, gm/cm^3$, is therefore more buoyant than oceanic material, and stands proud of the ocean floors. Such crust may be composed of as much as 7 km of sedimentary and volcanic rocks, underlain by >30 km of granodiorite and metamorphic rocks. The greater thickness of the less dense rocks compared with the denser oceanic crust is nature's way of balancing out the mass column to the centre of the Earth.

Mercury
Mercury's iron-rich core is roughly 3600 km in diameter, too small for complete differentiation to have taken place. Because the planet has been discovered to have an intrinsic magnetic field, this means that the core is at least partially molten, thereby generating a dynamo, in much the same way as the Earth. Calculations suggest that the mantle layer is relatively thin, about 200–250 km thick, and is overlain by a 500 km-thick silicate crust (Figure 4.6).

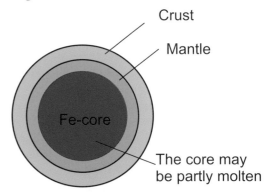

Figure 4.6 The internal structure of Mercury. The silicate mantle is relatively thin compared to that of the Earth.

MESSENGER's X-ray spectrometer has shown that the older, more heavily cratered terrain has higher ratios of magnesium to silicon, sulphur to silicon, and calcium to silicon, but lower ratios of aluminium to silicon than the smoother northern plains. There appear to be two clusters of data points for the older terrain: one with an Mg/Si ratio of 0.75 and an Al/Si ratio of 0.15, and another with Mg/Si and Al/Si ratios of 0.45 and 0.3, respectively. This suggests that the smooth plains material erupted from a magma source that was chemically different from the source of the material in the older regions. The high levels of sulphur – roughly ten times that found on Earth – set Mercury apart from the other terrestrial planets. The high Mg values suggest that high-temperature komatiite lavas contribute widely to the planet's crust. The relatively low Al and Ca concentrations and the high Mg content of Mercury's surface indicate that it is quite unlike the Moon's anorthositic highland crust.

The same spacecraft also showed that not only does water ice exist on Mercury, but in many places this ice appears to be covered in a 10 cm-thick layer of soot-like organic material. Despite being so close to the Sun, the surface is in permanent shadows within craters at the planet's poles. Here the temperature probably doesn't rise much above 170°C, allowing water ice to exist more or less indefinitely.

it is unable to convect at the present time. The mantle, presumed to be of similar composition to that of Earth, is also of similar dimensions.

A data set for Venus's crust was obtained by the Russian Venera and Vega landers, which analyzed major elements via XRF on rock powders, but lack data for Na and Cr. K, Th, and U were analyzed by gamma-ray spectrometry. The abundances of Fe, Mn, and Mg in the Venera and Vega basalts are comparable to those in Earth basalts, and suggest similar mantle compositions and core sizes (table 4.2).

Table 4.2 Chemistry at the Venera and Vega landing sites in wt %.

	Venera 13	Venera 14	Vega 2
SiO_2	45.1±6.0	48.7±7.2	45.6±3.2
TiO_2	1.6±0.9	0.2±0.2	0.2±0.1
Al_2O_3	15.8±6.0	17.9±5.2	16.0±3.6
SO_3	1.6±2.0	0.35±0.6	4.7±1.5
K_2O	4.0±1.2	0.2±0.14	0.1±0.08
CaO	7.1±2.0	10.3±2.4	7.5±0.7
Na_2O	n.d	n.d.	2.0
MnO	0.2±0.2	0.16±0.16	0.14±0.12
Cl	<0.3	<0.4	<0.3
Cu	n.d	n.d.	<0.3
Zn	n.d.	n.d.	<0.2
As, Se, Br	n.d.	n.d.	<0.08
Sr, Y, Zr, Nb, Mo	n.d.	n.d.	<0.1
Pb	n.d	n.d.	<0.3
Total	96.1%	95.8%	95.4%

Venus

It is assumed that Venus's core is of similar dimensions to that of Earth. However, because there is no measurable magnetic field, it is likely that the interior pressure at core depth is too great to allow it to crystallize, therefore

Mars

Modern measurements indicate that Mars has a completely molten metallic core with a diameter of c.3600 km. While being predominantly composed of metallic iron, there is a

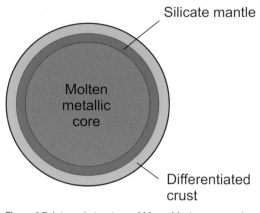

Silicate mantle

Molten metallic core

Differentiated crust

Figure 4.7 Internal structure of Mars. Modern research indicates that Mars has a molten iron-rich core.

significant proportion (16 wt %) of sulphur. The overlying silicate mantle is about 100 km thick with a differentiated 100-km-thick crust with a density of between 2.95 and 3.15 gm/cm^3. Analysis of meteorites believed to originate on Mars (SNCs), indicates that the Martian mantle is twice as rich in iron as Earth's (Figure 4.7).

Most of what we know about the elemental composition of Mars comes from orbiting spacecraft and landers. The two Mars exploration rovers currently operating each carry an Alpha Particle X-ray Spectrometer (APXS), a thermal emission spectrometer (Mini-TES), and Mössbauer spectrometer to identify minerals on the surface. Currently the Curiosity rover is sampling rocks and soils in the region of Mount Sharp, the central peak on the floor of Gale Crater, in Aeolis Palus.

The mineral olivine occurs all over the planet, but some of the largest concentrations are in Nili Fossae and Ganges Chasma, areas of Noachian-aged rocks. Since this mineral readily breaks down in the presence

of water, it can be assumed that liquid water has not been abundant since the rocks formed. Pyroxenes are also widespread, both the orthorhombic (low-Ca) and monoclinic (high-Ca) species being present. The high-Ca types are more typical of the large volcanic shields, while the low-Ca types more typical of the highland rocks. Plagioclase also is very widespread. Thus we have all three of the essential minerals typical of the rock basalt.

Between 1997 and 2006, the Thermal Emission Spectrometer (TES) on the Mars Global Surveyor (MGS) spacecraft mapped the global mineral composition of the planet, identifying two global-scale volcanic units on Mars: (i) Noachian-aged highlands in which basalts contain unaltered plagioclase and clinopyroxene, and (ii) younger plains north of the dichotomy boundary in which basalts are richer in SiO_2. The lavas of the second type have been interpreted as andesites or basaltic andesites, indicating that the lavas in the northern plains originated from more chemically evolved, volatile-rich magmas; however, the differences may be more apparent than real, simply representing more weathered samples of similar rocks.

In early 2004, the American MER Spirit rover landed in Gusev crater and MER Opportunity rover landed in Meridiani Planum, carrying instruments able to directly identify various minerals in Martian soils. True intermediate and felsic rocks are present, but exposures are uncommon. Instruments on-board the Mars Odyssey spacecraft have identified high silica rocks in Syrtis Major and near the rim of the crater Antoniadi. Geophysical evidence suggests that the bulk of the Martian crust may actually consist of basaltic andesite or andesite, the andesitic crust being hidden by the overlying basaltic lavas.

If the soils at Gusev crater and Meridiani Planum are representative of the Martian crust as a whole, then relatively unweathered basalts comprise about three-quarters (by weight) of the surface of Mars and older, aqueously-altered rocks comprise the remainder. Rocks studied by Spirit Rover in Gusev crater can be classified on the basis of their mineralogical proportions, as picritic basalts. These rocks are similar to ancient terrestrial rocks called basaltic **komatiites**.

In March 2013, NASA's Curiosity rover found evidence of mineral hydration, likely hydrated calcium sulphate, in several rock samples including the broken fragments of 'Tintina' rock and 'Sutton Inlier' rock as well as in veins and nodules in other rocks like 'Knorr' and 'Wernicke'. Analysis using the rover's DAN instrument provided evidence of subsurface water, amounting to as much as 4%, down to a depth of 60 cm.

Much of the Martian surface is deeply covered by dust as fine as talcum powder. The global predominance of dust obscures the underlying bedrock, making spectroscopic identification of primary minerals impossible from orbit over large areas. The red/orange appearance of the dust is caused by iron oxide and the mineral goethite. The global dust cover and the presence of other wind-blown sediments have made soil compositions remarkably uniform across the Martian surface. Analysis of soil samples from the Viking landers in 1976, Pathfinder, and the Mars Exploration rovers show nearly identical mineral compositions from widely separated locations around the planet. The soils consist of finely comminuted basaltic rock fragments

and are highly enriched in sulphur and chlorine, probably derived from volcanic gas emissions.

In 2008, the Compact Reconnaissance Imaging Spectrometer for Mars (CRISM), a visible-infrared spectrometer on-board Mars Reconnaissance Orbiter, recorded the spectral signature of the clay mineral smectite in the rim around Endeavour. The most recent Mars mission saw the rover Curiosity landing in Gale Crater on Mars in August 2012. In October of the same year, it performed the first X-ray diffraction analysis of Martian soil. The results revealed the presence of feldspar, pyroxenes and olivine, and suggested that the Martian soil in the sample was similar to weathered basaltic soils such as those found in Hawaii. The major element make-up of the rock 'Esperance' is shown in Figure 4.8.

The first powder rock sample ever collected on Mars had traces of sulphur, nitrogen, hydrogen, oxygen, phosphorus, and carbon. About 20 to 30% of the sample was

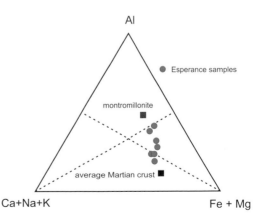

Figure 4.8 Triangular diagram plot for Esperance rock samples, together with average Martian crust and montmorillonite.

made of clay minerals. A subsequent sample was high in elements consistent with the mineral feldspar, and low in magnesium and iron. Curiosity also imaged pebbles that were apparently consistent with deposition in and by water.

Analyses of rocks from the Moon, Venus, Mars and the asteroid Vesta are shown on a standard SiO_2 vs Na_2O+K_2O diagram which also shows the fields for terrestrial volcanic rocks for comparison. As can be seen, the majority fall within the basalt to basaltic–andesite fields, but there are some more alkaline differentiates in the Martian group, as well as from Venus. Because of its much greater degree of differentiation, the Earth, with its development of oceanic and continental crust, shows a far greater range of rock chemistry.

Outer planets and moons

All of the gas giants, with the possible exception of Neptune, appear to have rocky cores, while the larger satellites of Jupiter also are expected to have iron-rich cores. The substantial magnetic field associated with Ganymede implies that this moon has a partially molten metal core. Many smaller bodies, such as the icy moons of the outer planets, are likely to have small silicate cores, but some may have been too small to differentiate completely, their interiors remaining as mixtures of rock and ice.

The mantles of the satellites of the gas giants may well behave in a similar manner to that of the Earth, but rather than being composed of silicates, are composed of ices, probably in different structural states dependent upon depth. In contrast to most silicate minerals, the melting temperature of ice falls with rising pressure (i.e. < 200 MPa) but then rises again. This means that there are likely to be sub-surface layers of water on moons such as Europa, Enceladus and Titan. These will be equivalent to the magma oceans that existed on the inner planets and Earth's Moon. Figure 4.9 shows the internal structures of the major planets.

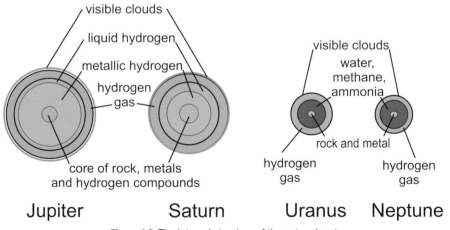

Figure 4.9 The internal structure of the outer planets.

5 Magnetic fields

The Sun's powerful magnetic field – the **heli-osphere** – generates a stream of high-energy charged particles that power outwards across space, enveloping all of the planets and their satellites. This gusty stream of solar charged particles travels at velocities of 200–900 kps in all directions and consists of roughly equal numbers of protons and electrons along with some alpha particles and heavier ions. This solar wind was responsible for stripping away the primary atmospheres of the youthful planets. Fluctuations in the strength of this field, together with the regularly changing cycle of sunspot activity, are known to have an effect on both climate and communications.

The interplanetary magnetic field

The magnetic field carried by the solar wind among the planets of the Solar System is known as the **interplanetary magnetic field** (IMF). This results from the influence of the Sun's rotating magnetic field on the plasma in the interplanetary medium. It rotates along with the Sun with a period of about 25 days, during which time the peaks and troughs of the skirt pass through the Earth's **magneto-sphere**, interacting with it. The outer limit of this field is termed the **heliopause**.

As the Sun rotates, its magnetic field twists into a Parker spiral, a form of Archimedean spiral, as it extends through the solar system (Figure 5.1). This spiral changes the shape of the Sun's magnetic field in the outer part of the solar system and also greatly amplifies its strength. It has a shape similar to a twirled ballerina skirt, and changes in shape through the solar cycle because the Sun's magnetic field reverses about every 11 years.

The plasma in the interplanetary medium is also responsible for the strength of the Sun's magnetic field at the orbit of the Earth. If space were a vacuum, then the Sun's magnetic **dipole** field, about 10^{-4} teslas at the surface of the Sun, would reduce with the inverse cube of the distance. However, satellite observations show that it is about 100 times greater than that. Theory predicts that the motion of a conducting fluid in a magnetic field induces electric currents, which in turn generates

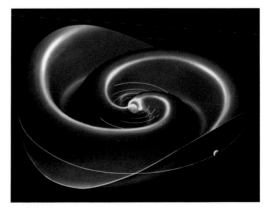

Figure 5.1 A Parker spiral. The direction of the Sun's magnetic field changes in the outer part of the Solar System and the spiralling motion increases its strength there.

magnetic fields, and in this respect it behaves like a magnetohydrodynamic dynamo.

NASA's Voyager probes recently passed through the field in the very outermost reaches of the Solar System and revealed some very interesting facts. Where the Voyagers are now, the folds of the skirt bunch up. When the field gets severely folded like this, interesting things can happen and the lines of magnetic force criss-cross, and 'reconnect'. (Magnetic reconnection is the same energetic process underlying solar flares.) The crowded folds of the skirt reorganize themselves, sometimes explosively, into foamy magnetic bubbles. This is what the Voyagers have been passing through. According to computer models, the bubbles are large, about 100 million miles wide, so it would take the speedy probes weeks to cross just one of them. Voyager 1 entered the 'foam-zone' around 2007, and Voyager 2 followed about a year later.

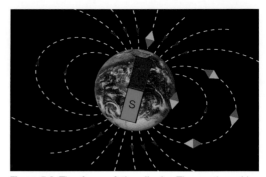

Figure 5.2 The form of the dipole. The north-seeking pole of a compass is attracted to the north pole of the dipole magnet.

Earth's magnetic field

The Earth has a magnetic field, and has had one presumably from the time the liquid core segregated. This has played the vital role of protecting the surface from most of the harmful radiation that would otherwise reach us from space. The Sun, as we have seen, also has such a field, but it is infinitely more powerful than Earth's and, on an almost permanent basis, interacts with and modifies our own. Massive explosions and surges in emissions from the Sun often play havoc with radio transmissions and also play a part in modifying Earth's climate. The 11-year sunspot cycle also affects climate, something known to the Chinese 3000 years ago.

The dominant structure of this field is called the dipole (Figure 5.2), which is roughly aligned with the planet's rotational axis. As is typical of a huge bar magnet, the field's primary magnetic flux is directed out of the core in the southern hemisphere and down towards the core in the northern hemisphere. Thus the field's magnetic north pole lies beneath the north geographical pole.

Research shows that the flux is not distributed evenly across the globe; most of the dipole's field overall intensity originates beneath North America, Siberia and the coast of Antarctica. These huge patches are probably part of a changing pattern of convection within the core. Evidence from the geological record shows that polarity reversals occur over short periods (4000 to 10 000 years). It would take the dipole at least 100 000 years to disappear on its own if the geodynamo were to shut down. Such a rapid change suggests that some kind of instability destroys the original polarity while generating a new one.

The solar wind varies in intensity dependent upon the amount of surface activity on the Sun. The Earth's magnetic field shields the terrestrial surface from much of this, for when the solar wind encounters Earth's magnetic field

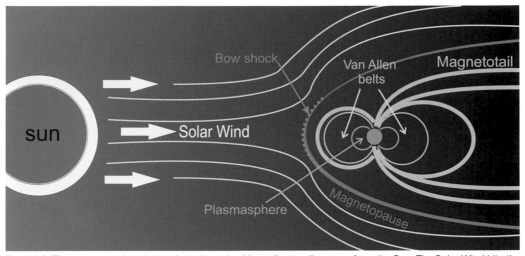

Figure 5.3 The magnetosphere is teardrop-shaped, with a tail extending away from the Sun. The Solar Wind hits the front of the magnetosphere at the Bow Shock.

it is deflected like water around the bow of a ship (Figure 5.3). The Earth's field is believed to be driven by a self-perpetuating dynamo within the liquid outer core.

Of huge importance to our understanding of the Earth, geologists discovered in the twentieth century that the Earth's magnetic field provided vital clues as to how new sea floor forms, thereby producing new oceanic crust, and how the continents and oceans have changed their positions with respect to one another. This is the study of **paleomagnetism**.

Origin of the magnetic field

Magnetic fields are produced by the motion of electrical charges; for example, the magnetic field of a bar magnet results from the motion of negatively charged electrons in the iron of the magnet. The origin of the Earth's magnetic field is not completely understood, but generally is believed to be associated with electrical currents produced by the coupling of convective motions and rotation in the Earth's spinning liquid metallic outer core. This mechanism is known as the *dynamo effect*.

A fundamental property of magnetic fields is that they exert forces on moving electrical charges. Thus, a magnetic field can trap charged particles such as electrons and protons as they are forced to execute a spiralling motion back and forth along the field lines. One of the first fruits of early space exploration was the discovery in the late 1950s that the Earth is surrounded by two regions of particularly high concentration of charged particles called the **Van Allen radiation belts** (*see* Figure. 5.3). The primary source of these charged particles is the solar wind. Charged particles trapped in the Earth's magnetic field are responsible for **aurorae**.

Three basic conditions are required to generate a planet's magnetic field:
1) A large volume of electrically-conducting fluid – i.e. a molten, iron-rich core.

2) A supply of energy to move the fluid. The energy driving the Earth's dynamo is part thermal and part chemical – both create buoyancy deep within the outer core. Thus iron convects upwards, cools and sinks again. Also, as fluid reaches the upper core boundary, it loses some heat to the overlying mantle. The liquid iron then cools and crystallizes, releasing latent heat that also contributes to thermal buoyancy.

3) For a self-sustaining dynamo we also need rotation. All planets do this. Earth's spinning provides this through the **Coriolis effect** by which, **in** the core, the Coriolis force deflects the upwelling fluid into corkscrew-like, helical paths.

Paleomagnetism

Study of the intensity and orientation of the Earth's magnetic field as preserved in the magnetic orientation of certain minerals found in rocks of all ages is called paleomagnetism. In the late 1920s, the French physicist Mercanton suggested that because today's magnetic field is close to the Earth's rotational axis, **continental drift** could be tested by establishing the magnetic characteristics of ancient rocks.

Rocks that crystallize from the molten state contain indicators of the ambient field at the time of their solidification. Thus some iron minerals are naturally magnetic and when they crystallize from a rock melt, they take on the polarity of the time and also have their tiny internal magnets aligned in the same direction as the prevailing field (Figure 5.4). The study of such fossilized magnetism indicates that the Earth's magnetic field frequently reverses itself (termed a **magnetic reversal**) or makes a good attempt at doing so (a **magnetic excursion**).

Paleomagnetism is possible because some of the minerals that make up rocks – notably

Figure 5.4 Thin section of a typical terrestrial basalt. The black grains are of the magnetic mineral, magnetite. When this crystallizes from the magma the tiny internal magnets (yellow arrows) take on the orientation of the ambient field.

the iron oxide magnetite, in combination with maghemite – become permanently magnetized parallel to the earth's magnetic field when the temperature falls below the Curie Point (620°C) and they crystallize. Geophysicists have been able to trace changes in the orientation of the Earth's magnetic field through geological time by carefully collecting rock specimens of different ages and determining the alignment of their magnetized particles.

Paleomagnetic studies of rocks and ocean sediment have demonstrated that the orientation of the Earth's magnetic field has frequently alternated over geologic time. Periods of 'normal' polarity, when the north-seeking end of the compass needle points toward the present north magnetic pole, have alternated with periods of 'reversed' polarity (when the north-seeking end of the compass needle points southward (Figure 5.5).

Over the past few years it has been discovered that new 'reversed flux' magnetic patches have formed under the east coast of North America and the Arctic. It has also been noted that the growth and pole-ward migration of these reversed flux patches accounts for the historical decline of the dipole. What is more, the primary geomagnetic field has lessened by nearly 10% since the 1830s, which is about 20 times faster than the field would decline naturally were it to lose its power source. Is this just a fluctuation or could another reversal be on its way?

There is another feature of magnetic minerals that has proved valuable to geologists: if a freely-suspended magnetic needle is held above various points on the Earth's surface, its inclination will not be the same everywhere. For instance, above either of the poles it will dip vertically; in contrast, above the Equator it will hang horizontally. For positions in between the two extremes the angle of its inclination will vary with the latitude. This **magnetic inclination**, as it is termed, can be measured in ancient rocks that contain magnetic minerals and, to cut a long story short, it has enabled geologists to establish at what latitudes continental blocks have resided at different times in the past. When the technique was originally discovered, it appeared that the magnetic pole had wandered around, since the magnetic data from different continental blocks did not coincide. However, since we know that the Earth's internal magnet does not change its position, the only deduction was that it was the continents that had moved – hence continental drift.

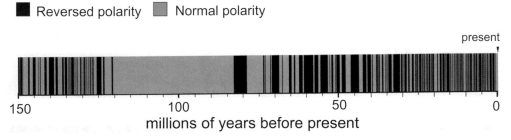

Figure 5.5 Periods of normal and reversed polarity during the past 150 million years.

The study of both aspects of rock magnetization has been one of the fundamental tools that has facilitated our understanding of plate tectonics and enabled us to gain at least a basic understanding of how the Earth generates new crust, and how the continents and oceans have changed their positions over the course of time.

Mercury

Mercury has a significant magnetic field which was first detected by Mariner 10, and fully confirmed by MESSENGER, which first went into Mercury orbit in 2011. The strength is about 1.1% of the Earth's field (about 350 gammas); it is dipolar, and the magnetic axis is virtually coincident with the axis of rotation. It is strong enough to deflect the solar wind, so that there is a small but very definite magnetosphere; there is a bow shock at about 1.5 Mercurian radii from the centre of the globe. In all probability it is generated by a dynamo effect, as with the Earth's field, and this means that Mercury must have an outer liquid core.

Venus

The very slow rotation of Venus means that despite being a near twin of the Earth in terms of size and bulk composition, it does not generate a comparable magnetic field. An extremely weak field has been measured; this is 100 000 times weaker than Earth's and can, to all intents and purposes, be ignored.

Mars

Mars's current magnetic field is very weak, with strengths of, at most, about 1.5 nanotesla. Earth's, by comparison, is about 50 microtesla, which is more than 3000 times stronger than Mars's field. Earth's magnetic field is supported by an internal dynamo, and while Mars must once have had a dynamo, which would have magnetized its rocks, for some reason the dynamo shut down. This could have been due to the relatively small mass of Mars, which meant that its core cooled quickly and solidified. For this reason any rocks that formed on Mars before the dynamo shut down are magnetized, while rocks that formed after the dynamo shut down, or were heated above their Curie point after the dynamo shut down, are not magnetized.

Mars has two very different hemispheres, the northern having been resurfaced largely by volcanic activity, while the southern is composed of elevated ancient cratered crust. The line between them is known as the planetary dichotomy. It has never been easy to understand why Mars's northern hemisphere has very little in the way of a detectable magnetic field, because evidence suggests that the dichotomy is a truly ancient feature that should have formed before the dynamo shut down.

Recently, a suite of papers published in *Nature* gave credence to the idea that the crustal dichotomy formed as a result of a truly gigantic, oblique impact that blasted away a massive chunk of the crust of the northern hemisphere, and that this happened very early in Mars's history. Furthermore, the papers showed that because the impact struck Mars a glancing blow, it would not have delivered enough heat to the planet to melt its whole crust. So the impact could reasonably leave Mars with a thin northern hemisphere crust and a thick southern hemisphere crust. But that still leaves geophysicists with the problem of explaining how the north didn't become magnetized, despite the fact that the dynamo was operating well after this putative giant impact.

In a fascinating *Science* paper, Sabine Stanley and her co-authors, using a neat computer simulation, showed that it is possible to have a dynamo operating inside Mars that only produces a dynamo in a single hemisphere. The way they made this happen is to suppose that instead of assuming a spherically symmetrical Mars, to make the northern hemisphere core-mantle boundary hotter than the southern hemisphere core-mantle boundary, a reasonable initial condition to impose if you very suddenly remove a huge amount of crust from that part of the planet with a massive impact.

Having run their simulation they produced two maps, the first of which shows a magnetic field simulated for a planet with no temperature difference at its core–mantle boundary. The second map shows their simulated Mars. This shows a strong magnetic field only in the southern hemisphere (Figure 5.6). They also propose that as a result of this situation, if ancient Mars had a magnetic field only in the southern hemisphere, then there would be no protective field to prevent the solar wind from stripping away the atmosphere from the north. However, atmosphere flows where it will, regardless of magnetic field lines; so atmosphere would flow from south to north, and continually be lost from the north. So you could still have Mars losing its atmosphere early in its history, even while it had an active dynamo.

Jupiter

Some 20000 times stronger than Earth's magnetic field, Jupiter's magnetic field creates a magnetosphere so large it begins to avert the solar wind almost 3 million kilometres before it reaches Jupiter. The magnetosphere extends so far past Jupiter it sweeps the solar wind as far as the orbit of Saturn. Like Earth's magnetosphere, many of the charged particles trapped in Jupiter's magnetosphere come from the solar wind; however, Jupiter has an extra source of particles that other planets do not have. Jupiter's volcanically active moon, Io, provides a substantial portion of charged particles to Jupiter's magnetosphere.

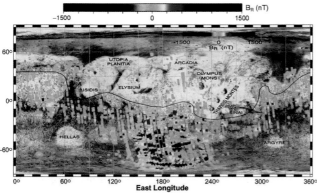

Figure 5.6 The strength of the magnetic field at the surface of Mars. Recent research has shown that it is possible to have a dynamo inside Mars that only operates in a single hemisphere. Courtesy NASA/HST.

The strength is 4.2 G at the Jovian equator and 10–14 G at the magnetic poles; by contrast, the strength of the Earth's magnetic field at the equator is a mere 0.3 G. With Jupiter, the magnetic axis is inclined to the rotational axis at an angle of 9.6°. The polarity is opposite to that of the Earth, so that if it were possible to use a magnetic compass on Jupiter the needle would point south. With regard to magnetic phenomena, distances from the centre of Jupiter are usually reckoned in terms of the planet's radius (RJ). The volcanic satellite Io, which has such a profound effect upon these phenomena, lies at a distance of 5.9 RJ or about 422 000 km.

The magnetic field is generated inside Jupiter, near the outer boundary of the shell of metallic hydrogen that lies below the atmosphere. The field is not truly symmetrical, but beyond a distance of a few RJ it more or less corresponds to a dipole. This magnetosphere is one of the largest features in the solar system. The plasma of electrically charged particles that exists in the magnetosphere is flattened into a large disk more than 4.8 million kilometres in diameter, is coupled to the magnetic field, and rotates around Jupiter. The Galilean satellites are located in the inner regions of the magnetosphere and are subjected to intense radiation bombardment.

The **magnetopause**, where the incoming solar wind particles are deflected, produces a bow shock at about 10 RJ ahead of the actual magnetopause. On the sunward side of Jupiter the field tends to be compressed; on the night side it is stretched out into a magnetotail which may be up to 650 million kilometres long, so that at times it can even engulf Saturn.

There are zones of intense radiation at least 10 000 times more powerful than the Van Allen zones that surround the Earth. Pioneer 10, the first space-probe to encounter these zones (in 1972), received a total of over 250 000 rads (a dose of 500 rads is fatal to humans!). It is now known that the Galilean satellite Io is connected to Jupiter by a very strong flux tube setting up a potential difference of 400 000 V. Molecules are sputtered off Io's surface by particles in the magnetosphere, producing a 'torus' tilted to Io's orbit, so that Io passes through it twice for each rotation of Jupiter.

Intense auroral activity has been imaged many times. The Hubble Space Telescope imaged these unusual light displays in more detail than ever before. Jupiter's auroras are linked to its volcanic moon Io. Io's volcanoes release particles, some of which become ionized, trapped by Jupiter's magnetic field, and rain down on the gas giant. The resulting auroral displays may be thousands of times brighter than any auroral display on Earth, and involve unusual spots. The above pictures show how the extended auroral emissions rotate with Jupiter, while the auroral spots stay synchronized to Io as it circles Jupiter (Figure 5.7).

Saturn

Saturn has a strong magnetic field that was first detected in 1979 from Pioneer 11. The strength of the field is around 600 times that of the Earth, though 30 times weaker than Jupiter's. Saturn is unique inasmuch as the magnetic axis and the rotational axis are almost coincident, so that the magnetic field is reasonably straightforward. The polarity is opposite to that of the Earth – that is, like Jupiter. The field deviates measurably from a simple dipole field; this manifests itself both in a north–south asymmetry and in a slightly

Figure 5.7 An image of Jupiter
obtained with NASA's Hubble
Space Telescope. This captured a
complete view of Jupiter's northern
and southern auroras. Courtesy
NASA/HST.

Jupiter Aurora
Hubble Space Telescope • STIS • WFPC2

PRC98-04 • ST ScI OPO • January 7, 1998 • J. Clarke (University of Michigan) and NASA

higher polar surface field than is predicted by a pure dipole model. The field seems to be slightly stronger in the north than in the south, and the centre of the field is displaced by about 2400 km northward along the planet's axis.

The magnetosphere is about one-fifth the size of that of Jupiter, the bow shock lying at a range of between 20 and 50 times the radius of the planet, and is about 2000 m thick. Depending on variations in the solar wind, the bow shock moves in and out to some extent; when the Cassini spacecraft was approaching Saturn, it crossed the bow shock no fewer than seven times.

Saturn's magnetosphere traps radiation belt particles, extending as far as the outer edge of the Main Ring system. However, at the outer edge of Ring A, the numbers of electrons fall off quickly. All of the inner satellites are embedded in the magnetosphere, and there are significant effects due to emissions from the active satellite Enceladus. There is also a well-defined magnetotail on the planet's night side.

Figure 5.8 In January and March 2009, NASA's Hubble Space Telescope imaged Saturn when its rings were edge-on, resulting in a unique view of symmetrical auroral displays at both poles. Courtesy NASA.

The outer boundary is somewhat variable, and is close to the orbit of Titan, so that Titan is sometimes just inside the magnetosphere and sometimes just outside it. Polar aurorae have been recorded by all fly-by missions, beginning with Pioneer 11, and from satellites such the IUE (International Ultra-violet Explorer) and the Hubble Space Telescope. Such aurorae can rise over 2000 km above the cloud tops (Figure 5.8).

Uranus

The Uranian magnetosphere is very similar to the terrestrial magnetosphere. However, the polarity is opposite to that of the Earth. There is a bow shock that extends to 23 Uranian radii and deflects the supersonic flow of the solar wind in front of the magnetosphere and an extensive magnetotail. The forward part of the magnetosphere extends to roughly 25 planetary radii and the bow shock to about 33 planetary radii.

The equatorial field strength at the equator is 0.25 G, as against 4.28 G for Jupiter (the value for Earth is 0.305 G). However, the magnetic axis is displaced from the rotational axis by 58.6°; neither does the magnetic axis pass through the centre of the globe, being offset by 8000 km. The planet's 'sideways' rotation twists the tail into the shape of a corkscrew, while the field strength in the northern hemisphere is appreciably higher than in the southern. In the case of Jupiter and Saturn, the magnetic fields are generated in their cores;

however, those of Uranus and Neptune may well originate at much shallower levels, in the global liquid.

Neptune

It had been assumed for many years that Neptune must have a magnetic field; however, proof was not forthcoming until the arrival of Voyager 2. This probe passed across the bow shock at 879000 km from the planet. The magnetic field itself was found to be weaker than those of the other giants; the field strength at the surface is 1.2 G in the southern hemisphere but only 0.06 G in the northern. Surprisingly, the inclination of the magnetic axis relative to the axis of rotation is 47°, so that in this respect Neptune is not unlike Uranus; moreover, the magnetic axis does not pass through the centre of the globe, but is displaced by 10000 km or 0.4 Neptune radii. This indicates that the dynamo currents must be closer to the surface than to the centre of the globe. The Neptunian magnetosphere is probably the most quiescent magnetosphere in the Solar System.

Outer planet moons

Jupiter, having such a powerful magnetic field, induces fields on some of its Galilean moons. Thus, induced fields have been recorded on both Europa and Callisto; however, at present there is no certainty in this respect with Ganymede, but it is suspected of having a weak intrinsic field. When NASA's Galileo probe passed Io, however, violent oscillations in the ambient field suggested that it may have an intrinsic field.

6 Planetary atmospheres

The atmospheres of the planets are significantly different from one another, some worlds having retained their primary atmosphere, others losing it and developing a secondary one at a later stage. The primary atmospheres were composed mostly of light gases that accreted during initial condensation from the solar nebula. Thus the gases are similar to the primordial mixture of gases found in the Sun and Jupiter, i.e. 94.2% H, 5.7% He and everything else, less that 0.1%.

More massive worlds were able to retain the lighter elements (e.g. H and He) due to their relatively high escape velocity, whereas less massive planets lost most of these and could only hold on to the heavier atoms and molecules (e.g. CO_2, H_2O). The original H and He has thus been lost from the inner planets, and they procured a secondary atmosphere by later outgassing or by input from incoming bodies.

Origin of planetary water

It is a long-held theory that much of the water currently existing on the inner planets derived from comets that came in during the latter stages of solar system evolution (Figure 6.1). Recently, however, this notion has been

Figure 6.1 This close-up look at Comet ISON is from NASA's Hubble Space Telescope. This image was captured on 10 April 2013, when the comet was at a distance of 570 million km from the Sun. This image was taken in visible light, the blue false colour being added to bring out details in the comet's structure. (Image credit: NASA, ESA, J.-Y. Li (Planetary Science Institute), and the Hubble Comet ISON Imaging Science Team.)

challenged, for several reasons. Looking at the ratio of hydrogen to its heavy isotope deuterium in frozen water (H_2O), scientists can get an idea of the relative distance from the Sun at which objects containing the water were formed. Objects that formed farther out should generally have higher deuterium content in their ice than objects that formed closer to the Sun, and objects that formed in the same regions should have similar hydrogen isotopic compositions. Therefore, by comparing the deuterium content of water in carbonaceous chondrites to the deuterium content of comets, it is possible to tell if they formed in similar reaches of the Solar System.

After analysis of 85 carbonaceous chondrites, it was shown that carbonaceous chondrites probably did not form in the same regions of the Solar System as comets because they have much lower deuterium content. If so, this result directly contradicts the two most prominent models for how the Solar System developed its current architecture. Rather than deriving from comets, it has been suggested that most of the volatile elements on Earth arrived in chondrite meteorites.

Onset of terrestrial volcanism and degassing

Once the Earth had developed its layered internal structure, the 'heat machine' provided sufficient energy for the onset of volcanic activity. Molten magma from the upper part of the mantle began to force its way through the thin, brittle crust, both thickening and strengthening it, and acting as an insulator for the heat trying to escape from the hot interior. This early crust was made predominantly from the rock **basalt**.

As the molten lava spread out over the surface, it brought with it volatiles from within

the Earth, gases such as water (H_2O), carbon dioxide (CO_2), hydrogen sulphide (H_2S) and sulphur dioxide (SO_2). As volcanism continued, so it brought more and more volatiles towards the surface of the planet (Figure 6.2). It was from this source that at least some of the secondary atmosphere of the Earth derived. It also was responsible for the generation of a hydrosphere.

Figure 6.2 Active gas and fumarole field at Kawah Mas, Central Java. Volcanic outgassing contributed to the development of Earth's secondary atmosphere. Photo: the author

Mercury

Because of its low escape velocity ($4.25\,km^{S-1}$) Mercury would not be expected to have much of an atmosphere. Indeed, the atmosphere is very tenuous, the actual ground density being 10^{-15} of a bar – equivalent to a good laboratory vacuum. It is more appropriate to call it an exosphere, and its total mass is probably around only 1000 kg. The main constituents are oxygen (42%), sodium (29%) and hydrogen (22%). The exosphere is not stable, atoms being continuously lost and replenished from various sources.

In 2008 and 2009, the MESSENGER probe discovered several different ions in the vicinity of Mercury, including H_2O^+ (ionized water vapour) and H_2S^+ (ionized hydrogen sulphide) and magnesium. It is believed that charged particles streaming outwards in the solar wind have a clever way of penetrating the planet's protective magnetic field. The planet has tornado-like magnetic vortices that let charged particles from the sun get through; these kick up atoms from the surface that replenish the planet's thin atmosphere.

Venus

The atmosphere of Venus is 96.5% by volume carbon dioxide. Most of the remaining 3.5% is nitrogen, with minute traces of carbon monoxide, helium, argon, sulphur dioxide, oxygen and water vapour and other gases such as krypton and xenon. The clouds are also rich in sulphuric acid. The troposphere extends to an altitude of 65km, above which the stratosphere and mesosphere extend to 95 km. The upper atmosphere reaches out to at least 400km.

The mass of Venus's air is about 90 times that of the Earth's. Whereas 90% of the Earth's atmosphere is within 10km of the surface, it is necessary to go to 50km to capture 90% of that of Venus. The clouds of Venus may extend from 50 to 70km and may be divided into three distinct layers. Beneath the clouds is a layer of haze down to about 30km, below which it is clear. Above the clouds there is a high-speed jet stream which blows from west to east at velocities of between 300–400km/h. This is fastest at the equator and slows toward the poles, often giving a 'V' type pattern in the visible cloud cover. At the surface there is almost no prevailing wind, with measured surface wind speeds typically less than 2 m/s (Figure 6.3).

Figure 6.3 These Galileo images of Venus show the cloud patterns at two different depths in the upper cloud layers. The large bluish image, taken through the violet filter, shows patterns at the very top of Venus' main sulphuric acid haze layer. The small red image, taken through a near infrared filter, shows the cloud patterns several miles below the visible cloud tops. Courtesy NASA/JPL.

The entire atmosphere can be best described as super-rotating. The winds decrease from $100\,m/s^{-1}$ at the cloud-tops to only $50\,m/s^{-1}$ at 50km, and only a few metres per second at the surface. The atmospheric pressure at the surface is about 90 times greater than that of the Earth's air at sea level – roughly equivalent to being under water on the sea floor of Earth at a depth of 9km. The heat-trapping effect of the carbon dioxide leads to a surface temperature of around 467°C.

Presumably early volcanic activity generated sulphur, which rose into the atmosphere, and because of the elevated temperatures this could not be locked into solid compounds on the surface as it was on the Earth. In the hugely dense atmosphere of the planet, sulphur would be volatile enough to evaporate and form sulphur dioxide compounds which would then remain airborne.

There is also evidence of a Hadley cell, which means that hot air rises at the equator and is transported pole-wards in the upper atmosphere. This produces a polar dipole that rotates around the pole itself. This polar vortex is bounded by a polar collar, a wide (*c.* 1000 km) shallow river of cold air which circulates around the pole at about 70 degrees.

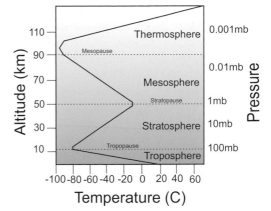

Figure 6.4 Vertical structure of the Earth's atmosphere.

Earth

The present atmosphere of the Earth is not the original one, for our current atmosphere is oxidizing, while the original atmosphere would have been reducing, probably containing no free oxygen. Free oxygen probably did not appear on Earth until around 2700 mya. The original atmosphere may have been similar in composition to the solar nebula but was lost to space, and replaced by compounds outgassed from the interior or from impacting planetesimals rich in volatile materials. The oxygen so characteristic of our atmosphere was almost all produced by photosynthesizing plants (cyanobacteria). The composition of the atmosphere is 79% nitrogen, 20% oxygen, and 1% other gases. Of all the planets, the Earth is the only one that has an atmosphere that can sustain life.

Layers of the Atmosphere

The atmosphere of the Earth may be divided into several distinct layers, as Figure 6.4 indicates:

1. The **troposphere** is where all weather takes place; it is the region of rising and falling parcels of air and contains 80% of the Earth's air. The air pressure at the top of the troposphere is only 10% of that at sea level (0.1 atmospheres).

2. The **tropopause** is the boundary between the troposphere and the stratosphere, varying in altitude from 8 km at the poles to 18 km at the equator.

3. The **stratosphere** and ozone Layer. Above the troposphere is the stratosphere, where air flow is predominantly horizontal. The thin ozone layer in the upper stratosphere has a high concentration of ozone. This layer is primarily responsible for absorbing the ultraviolet radiation from the Sun.

4. The **mesosphere** and **ionosphere**. The mesosphere lies above the stratosphere, and above that is the ionosphere, where many atoms are ionized. The latter is very thin, but it is where aurorae take place, and is also responsible for absorbing the most energetic photons from the Sun, and for

reflecting radio waves, thereby facilitating long-distance radio communication. The structure of the ionosphere is strongly influenced by the solar wind, which is in turn governed by the level of solar activity.

5. The **exosphere,** is a very tenuous region that extends outward until it interacts with the solar wind. Solar storms compress the exosphere such that when the sun is tranquil, the exosphere extends further outward. Its top ranges from 1000 km to 10 000 km above the surface, then merges imperceptibly with interplanetary space.

The Earth also is unique in having a voluminous hydrosphere, and one of the very important aspects of Earth's climate is the coupling of the atmosphere with the hydrosphere. The deeper waters of the ocean appear to comprise a global circulation pattern known as the Global Conveyor Belt, which has a major effect on the climate regime (Figure 6.5).

Figure 6.5 The global ocean conveyor that takes cold salty water sinking in the North Atlantic along the bottom of the ocean into the Pacific and back as the warm North Atlantic Drift. Courtesy SRH/NOAA.

Mars

The main constituent of the atmosphere is carbon dioxide, which accounts for more than 95% of the total; nitrogen accounts for 2.7% and argon for 1.6%, with traces of free oxygen, carbon monoxide, water and methane, for a mean molar mass of 43.34 g/mol. The highest atmospheric pressure so far measured is 890 pascals (8.9 mbar), on the floor of the deep **impact basin** Hellas, while the pressure at the top of Olympus Mons is below 300 pascals (3 mbar). The atmosphere is quite dusty, giving the Martian sky a light brown or orange colour when seen from the surface.

The atmospheric pressure is at present too low for liquid water to exist on the surface, but there is little doubt that water did once exist, for imagery shows clear evidence of old riverbeds and of episodes of major flooding. Most of this water now appears to be locked in a brecciated permafrost layer beneath the surface.

Mars may well go through very marked climatic changes, similar to those induced by Earth's precessional cycles. The effects of the changing axial inclination, which varies between 14.9° and 35.5° over a cycle of 51 000 years, and the changing orbital eccentricity, which ranges from 0.004 to 0.141 in a cycle of 90 000 years, must have a significant effect on climate.

Clouds are common (Figure 6.8). Some are due to ice crystals, such as the lee clouds formed downwind of major landscape obstacles, while wave clouds are seen at the edges of the polar caps. There are also fogs, usually seen in low-lying areas near dawn or dusk, ground hazes due to dust.

Temperatures in the troposphere decrease with altitude, as on Earth; the lapse-rate is around 1.5°/km up to the tropopause, at a height of about 40 km, above which the temperature remains fairly constant at 130°C. The ionosphere extends out to several hundred kilometres, and is really a collisionless gas.

Table 6.1 Composition of the Martian atmosphere at the surface (ppm).

Carbon dioxide	CO_2	95.32%
Nitrogen	N_2	2.7%
Argon	^{40}Ar	1.6%
Oxygen	O_2	0.03%
Carbon monoxide	CO	0.07%
Water vapour	H_2O	0.03% (variable)
Neon	Ne	2.5 ppm
Krypton	Kr	0.3 ppm
Xenon	Xe	0.08 ppm
Ozone	O_3	0.03 ppm

Trace amounts of methane (CH_4) were first reported in Mars's atmosphere by a team at the NASA Goddard Space Flight Centre; later, in 2004, the *Mars Express* Orbiter and ground-based observations confirmed this, with a mole fraction of about 10 nmol/mol.

Results from Curiosity

Curiosity was launched from Cape Canaveral on November 26, 2011, and successfully landed on Aeolis Palus in Gale Crater in August 2012. It has measured the abundances of different gases and different isotopes in several samples of the Martian atmosphere. It checked ratios of heavier to lighter isotopes of carbon and oxygen in the carbon dioxide that makes up most of the planet's atmosphere. Heavy isotopes of carbon and oxygen are both enriched in today's thin Martian atmosphere compared with the proportions in the raw material that formed Mars, as deduced from proportions in the Sun and other parts of the Solar System.

This provides not only supportive evidence for the loss of much of the planet's original atmosphere, but also a clue to how the loss occurred. The enrichment of heavier isotopes measured in the dominant CO_2 gas points to a process of loss from the top of the atmosphere, favouring loss of lighter isotopes rather than a process of the lower atmosphere interacting with the ground. Curiosity also measured the same pattern in isotopes of hydrogen, as well as carbon and oxygen, consistent with a loss of a substantial fraction of Mars's original atmosphere.

The Curiosity team believe that a catastrophic event must have torn the atmosphere apart 4 Gya. Some scientists consider this may have been major volcanic eruptions, while others suggest a massive collision stripped the atmosphere away. The results also show that reservoirs of carbon dioxide and water were established after this catastrophic event, and have remained little changed since. Whether or not life managed to gain a hold while the atmospheric composition was different, remains an area of speculation.

Jupiter

The atmosphere of Jupiter is the largest in the planetary system. The two main constituents are molecular hydrogen (H_2) and helium (He). The helium abundance is 0.157 ± 0.0036 relative to molecular hydrogen by number of molecules, and its mass fraction is 0.234 ± 0.005, which is slightly lower than the Solar System's primordial value. It also contains compounds such as water, methane (CH_4), hydrogen sulphide (H_2S), ammonia (NH_3) and phosphine (PH_3). Their abundances in the deep troposphere imply that the atmosphere of Jupiter is enriched in the elements carbon, nitrogen, sulphur and possibly oxygen by a factor of 2–4 relative to the Sun. The noble gases argon, krypton and xenon appear also to be enriched relative to solar abundances, while neon is scarcer.

Preliminary analysis of early data returned by NASA's Galileo probe provided a series of startling discoveries. Probe instruments found the entry region of Jupiter to be drier than anticipated, and the amount of helium measured was about one-half of that expected.

The atmosphere lacks a clear lower boundary and gradually merges with the liquid interior of the planet. The lowest layer, the troposphere, has a complicated system of clouds and hazes, comprising layers of ammonia, ammonium hydrosulphide and water. The upper ammonia clouds visible at Jupiter's surface are organized in zonal bands parallel to the equator and are bounded by powerful zonal atmospheric jets. The bands alternate in colour: the dark bands are called belts, while light ones are called zones. Zones, which are colder than belts, correspond to upwellings, while belts mark descending air (Figure 6.6).

The tropopause is approximately 50 km above the visible clouds where the pressure and temperature are about 10 pascals (0.1 bar) and 110 K. In the stratosphere, the temperatures rise to about 200 K at the transition into the thermosphere, at an altitude and pressure of around 320 km and 1 µbar. In the **thermosphere**, temperatures continue to rise, eventually reaching 1000 K at about 1000 km, where pressure is about 1 nbar.

In the equatorial region, the Jovian winds blow from west to east at about $100 \, \mathrm{m \, s^{-1}}$ relative to the core, reaching maximum

Figure 6.6 Jupiter imaged by Damian Peach on September 12 2010, when Jupiter was close to opposition (PIA14410). Note the zonal cloud belts and the 'Great Red Spot'. Jupiter's moons, Io and Ganymede, can also be seen. Courtesy NASA/Damian Peach.

speeds 6° to 7° north and south of the equator. In the northern hemisphere the east wind decreases with increasing latitude, until at 18° N the clouds are rotating westward at 25 m s⁻¹. North of this, the wind speed declines to zero, and then changes in an eastward direction, reaching a maximum of 170 m s⁻¹ at 24° N.

In the southern hemisphere conditions are not quite the same, due probably to the presence of the Great Red Spot, but in both hemispheres there is an alternating pattern of eastward and westward jet streams, which mark the boundaries of the visible belts. The coloured ovals and spots circulate cyclonically or anticyclonically as if revolving between the jet streams.

The atmosphere shows a wide range of active phenomena, including vortices that reveal themselves as large red, white or brown ovals. The largest two spots are the Great Red Spot and Oval BA, which is also red (Figure 6.7). These two and most of the other large spots are anticyclonic. Smaller anticyclones tend to be white. Vortices are thought to be relatively shallow structures with depths not exceeding several hundred kilometres. Located in the southern hemisphere, the GRS is the largest known vortex in the Solar System. It could engulf several Earths and has existed for at least 300 years. Oval BA, south of GRS, is a red spot a third the size of GRS, which formed in 2000 from the merging of three white ovals.

Figure 6.7 This Voyager 2 image shows the region of Jupiter extending from the equator to the southern polar latitudes in the neighbourhood of the Great Red Spot. A white oval is situated south of the Great Red Spot. Courtesy NASA/JPL.

The planet experiences powerful storms, always accompanied by lightning. The storms are a result of moist convection in the atmosphere, and are connected to the evaporation and condensation of water. They are sites of strong upward motion of the air, which leads to the formation of bright and dense clouds. The storms form mainly in belt regions.

Table 6.2 Elemental abundances relative to hydrogen in Jupiter and Sun.

Element	Sun	Jupiter/Sun
He/H	0.0975	0.807 ± 0.02
Ne/H	1.23×10^{-4}	0.10 ± 0.01
Ar/H	3.62×10^{-6}	2.5 ± 0.5
Kr/H	1.61×10^{-9}	2.7 ± 0.5
Xe/H	1.68×10^{-10}	2.6 ± 0.5
C/H	3.62×10^{-4}	2.9 ± 0.5
N/H	1.12×10^{-4}	3.6 ± 0.5 (8 bar)
		3.2 ± 1.4 (9–12 bar)
O/H	8.51×10^{-4}	00.19–0.58 (19 bar).
		0.033 ± 0.015 (12 bar)
P/H	3.73×10^{-7}	0.82
S/H	1.62×10^{-5}	2.5 ± 0.15

Saturn

The atmosphere of Saturn is made up of approximately 75% hydrogen and 25% helium, with trace amounts of ammonia, acetylene, ethane, propane, phosphine and methane. The outer atmosphere contains 96.3% molecular hydrogen and 3.25% helium. The proportion of helium is significantly deficient compared to its abundance in the Sun. The upper clouds are composed of ammonia crystals, while the lower level clouds appear to consist of either ammonium hydrosulphide (NH_4SH) or water.

The planet has some of the fastest winds in the Solar System. As the Voyager spacecraft approached Saturn, it recorded winds as fast as 1800 km/hour at the planet's equator.

Large white storms can form within the bands that circle the planet, but unlike Jupiter, these storms only last a few months and are absorbed into the atmosphere again.

The part of Saturn that we can see is the visible cloud deck (Figure 6.8). The clouds are made of ammonia, and sit about 100 km below the top of Saturn's troposphere, where temperatures decline to –250°C. Below this is a lower cloud deck made of ammonium hydrosulphide clouds. Here the temperature is only –70°C. The lowest cloud deck is made of water clouds, and located about 130 km below the tropopause. Temperatures here are at the freezing point of water.

As is the case with Jupiter, the cloud structures on Saturn rotate at different rates depending on latitude, and multiple rotation periods have been assigned to various regions. Thus System I has a period of 10 h 14 min 00 s

Figure 6.8 A stunning view of clouds in Saturn's southern hemisphere, images by NASA's Cassini spacecraft (PIA 08952). This false colour image suggests movement within bands of atmosphere. Courtesy JPL/NASA.

and encompasses the Equatorial Zone, which extends from the northern edge of the South Equatorial Belt to the southern edge of the North Equatorial Belt. All other Saturnian latitudes have been assigned a rotation period of 10h 38 min 25.4 s, which is System II. System III, based on radio emissions from the planet in the period of the Voyager flybys, has a period of 10h 39 min 22.4 s.

Saturn occasionally exhibits long-lived ovals and other features. In 1990, the Hubble Space Telescope imaged an enormous white cloud near Saturn's equator; this was not present during the Voyager encounters and in 1994, another, smaller storm was observed.

Uranus

Uranus has a relatively bland appearance, lacking the broad colourful bands and large clouds prevalent on Jupiter and Saturn. The atmosphere is calm compared to those of other giant planets, with only a limited number of small bright clouds at middle latitudes in both hemispheres and one Dark Spot, having been observed since 1986. However, the atmosphere has rather strong zonal winds blowing in the retrograde direction near the equator, but switching to the prograde direction poleward of ±20° latitude. The wind speeds are from −50 to −100 m/s at the equator increasing up to 240 m/s near 50° latitude.

The atmosphere is composed primarily of hydrogen and helium. At depth it is significantly enriched in volatiles, i.e. ices such as water, ammonia and methane. The opposite is true for the upper atmosphere, which contains very few gases heavier than hydrogen and helium due to its low temperature. Uranus's atmosphere is the coldest of all the planets, with a temperature reaching as low as 49 K.

The troposphere is the densest part of the atmosphere and holds almost all of the mass. It hosts four cloud layers: methane clouds at about 120 pascals (1.2 bar), hydrogen sulphide and ammonia clouds at 300–1000 pascals (3–10 bars), ammonium hydrosulphide clouds at 2000-4000 pascals (20–40 bars), and finally water clouds below 5000 pascals (50 bars). Only the upper two cloud layers have been observed directly. The planet's zonal winds are remarkably fast, with speeds up to 240 m/s. The temperature falls from about 320 K at the base at −300 km to 53 K at 50 km.

The overlying stratosphere shows a temperature range from 53 K in the tropopause to between 800 and 850 K at the base thermosphere. The heating of the stratosphere is caused by the downward heat conduction from the hot thermosphere as well as by absorption of solar UV and IR radiation by methane and the complex hydrocarbons formed as a result of methane photolysis.

Hydrocarbons heavier than methane are present in a relatively narrow layer between 160 and 320 km in altitude, corresponding to the pressure range from 1000 to 10 pascals (10–0.1 mbar) and temperatures from 100 to 130 K. The most abundant stratospheric hydrocarbons after methane are acetylene and ethane. In addition to hydrocarbons, the stratosphere contains carbon monoxide, as well as traces of water vapour and carbon dioxide. The Uranian thermosphere (exosphere) extends outwards for thousands of kilometres, and has a uniform temperature of around 800 to 850 K.

Neptune

Neptune's atmosphere is notably more interesting than that of Uranus with respect to its visible weather patterns. Thus, at the time of the 1989 Voyager 2 flyby, the planet's southern hemisphere possessed a Great Dark Spot comparable to the Great Red Spot on Jupiter (Figure 6.9). These weather patterns are driven by the strongest sustained winds of any planet in the Solar System, with recorded wind speeds as high as 2100 km/h.

Because of its great distance from the Sun, Neptune's outer atmosphere is extremely cold, with temperatures at its cloud tops approaching 55 K (–218°C). At high altitudes, the atmospheric composition is 80% hydrogen and 19% helium, with trace amounts of methane. As with Uranus, this absorption of red light by the atmospheric methane is part of what gives Neptune its blue hue.

The atmosphere is subdivided into two main regions; the lower troposphere, where temperature decreases with altitude, and the stratosphere, where temperature increases with altitude. The boundary between the two, the tropopause, occurs at a pressure of 10 kPa (0.1 bars). The stratosphere then gives way to the thermosphere at a pressure lower than 1 to 10 Pa (10^{-5} to 10^{-4} microbars) and gradually grades into the exosphere.

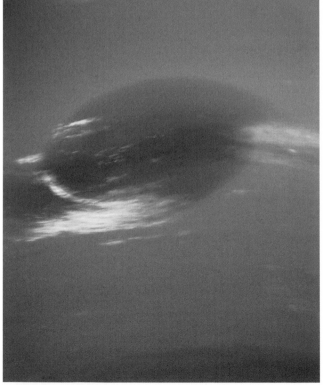

Figure 6.9 The great dark spot on Neptune. PIA00052. Courtesy JPL/NASA.

7 Geological development of the Inner Planets

Each of the inner planets heated up as accretion proceeded. Differentiation of metal and silicate phases then occurred, and each planet developed a silicate crust that was frequently punctured by magma escaping from inside and also impacted by bodies incoming from space. Thus these early crusts thickened, but were also modified by cratering. This impact record is seen to magnificent effect on Earth's Moon and on Mercury (Figure 7.1).

The ancient cratered crusts, or large areas of them, have survived on the Moon and

Figure 7.1 MESSENGER image of the multi-ringed basin, Michelangelo (PIA16759). Courtesy NASA/ Johns Hopkins University Applied Physics Laboratory/ Carnegie Institution of Washington.

Mercury, despite later modification, and on Mars, even with its greater degree of resurfacing. Major modifications on both Venus and the Earth have removed evidence of their primordial skins. The Moon's ancient crust is composed of **anorthosite**, a feldspar **cumulate** that crystallized 4.4 bya from a molten magma ocean rich in silica, aluminium and calcium. This bears the imprint of a period of intense bombardment by meteoroids that ended about 4 Gya. Within the inner Solar System it is the best preserved of the ancient surfaces we can study.

Early Earth

Once radioactive elements such as U, Pb, Sr, Nd became locked into the minerals found in near-surface rocks, they commenced to decay and an immense amount of heat was released. To this day, it is the long-lived radionuclides that provide a major source of internal heat. Their concentration in the outer layers of the Earth may have meant that an ocean of molten rock, a magma ocean, encased our planet. Once the dense, hot and partially molten terrestrial core formed, it became surrounded by the largely silicate mantle layer. Geologists believe this took around 200 Ma to achieve.

In due course, molten magma from the upper part of the mantle began to force its way through the thin, brittle crust, both thickening and strengthening it, and, by the way, acting as an insulator. This early crust was made from basalt. This is the rock that comprises the early crusts of all of the inner planets and is also important on Earth's Moon. It is by far the commonest rock in the Solar System.

Mercury

Mercury has a high density of 5.44 g/cm^3 which indicates the planet to be 60 to 70% by weight metal, and 30% silicate. This suggests a core radius of 85% of the planet radius and a core volume of 50% of the planet's volume. Mercury appears to have a solid silicate crust and mantle 600 km thick overlying a solid, iron sulphide outer core layer, a deeper liquid core layer, and possibly a solid inner core. The total core diameter is thought to be 3600 km. Although believed to be thin (100–300 km), the crust is an effective insulator that keeps the planet's outer core liquid and generates the observed magnetic field.

It has been suggested that an early major impact removed the majority of the planet's original mantle layer; however, when NASA's MESSENGER spacecraft went into orbit in March 2011, it detected considerable sulphur and potassium in the crust. Many planetary scientists believe that this rules out the giant-impact scenario; an energetic collision, they suggest, would have melted the planet and released such volatile constituents.

The oldest surfaces are peppered by a huge number of impact craters; crater counting techniques suggest an age for this of 3–4 Ga. Craters on such surfaces are >15 km in diameter. This crust is presumed to consist of brecciated basaltic material. Surprisingly, MESSENGER'S reflectance spectra data indicate low amounts of iron in the surface layer, but considerable amounts of sulphur. This is odd for primary rocks of early planetary crusts.

One of the most impressive landscape features is the Caloris Basin, the largest of several such basins; it has a diameter of 1550 km and an extensive ejecta blanket. Between the basins and larger craters are extensive smoother plains presumably created from ancient lava flows. Also very conspicuous are high scarps 100 m to 2 km high, which extend for thousands of kilometres. These formed in response to shrinking of the planet's core, the diameter of which may have decreased by at least 1.5 km (Figure 7.2).

MESSENGER data revealed that the spread in elevations is considerably smaller than those of Mars or the Moon. The most prominent feature is an extensive area of lowlands at high northern latitudes composed of volcanic plains. Within this lowland region is a broad topographic rise that formed after the volcanic plains were emplaced. At mid-latitudes, the interior plains of the Caloris impact basin have been modified so that part of the basin floor now stands higher than the rim. The elevated portion appears to be part of a quasi-linear rise that extends for approximately half the planetary circumference at mid-latitudes. These features imply that large-scale changes to Mercury's topography occurred after the era of impact basin formation and emplacement of volcanic plains had ended.

Venus

Most of what we know about the geology of Venus derives from the highly successful Magellan mission, which began imaging Venus in 1990. According to data from the earlier Pioneer altimeters, nearly 51% of the surface

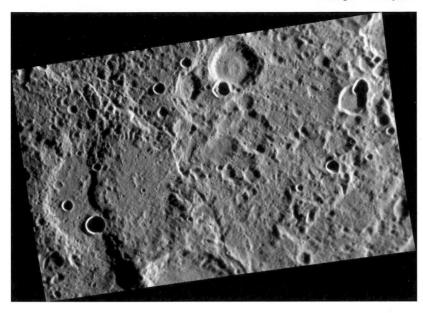

Figure 7.2 Mercury showing an impressive lobate scarp named Vostok Rupes. This was first seen by Mariner 10. Courtesy NASA/ Johns Hopkins University Applied Physics Laboratory/ Carnegie Institution of Washington.

is located within 500 metres of the median radius (6052 km), while only 2% of the surface is located at elevations greater than 2 km from this datum. Magellan's altimeter confirmed the general character of the landscape, its data showing that 80% of the topography is within 1 km of the median radius.

It also revealed far fewer impact craters on its surface than characterized Mercury, plus large numbers of faults and several prominent highland regions, the most important being Aphrodite Terra, Ishtar Terra, and Lada Terra, as well as the regions Beta Regio, Phoebe Regio and Themis Regio (Figure 7.3). Several of these returned complex radar signatures. Large expanses of the surface that gave dark radar signatures are built from volcanic plains with a low incidence of impact craters and many volcanic features. Thus Venus can be thought of in terms of two principal landscape types: *volcanic plains* (85%) and *deformed highlands* (15%). The former are covered in all manner of volcanic constructs, from very small shields to large domes and major volcanic shields. Individual and very extensive lava flows are discernible and have the typical appearance of fluid basalt.

Not surprisingly, therefore, basaltic rocks were identified as building the plains in Phoebe Regio by the Russian Venera 13 and 14 probes (*see* table 7.0). At the Vega lander sites on the border of Aphrodite Terra, measurements of radioactive elements indicated rocks akin to terrestrial gabbro. However, it is unlikely that any of these extensive plains are representative of the primeval crust of Venus.

The most striking elevated regions are undoubtedly the mountain belts that surround Lakshmi Planum: Maxwell Montes (11 km), Akna Montes (7 km) and Freya Montes (7 km).

Planetary Radius (km)

6048 6050 6052 6054 6056 6058 6060 6062

Figure 7.3 Magellan hemispheric view of Venus, showing region of Aphrodite Terra (left) and Beta Regio (right). PIA45389. Courtesy of NASA/JPL/USGS.

The more important of the deformed highland regions are Aphrodite Terra – which extends for 23 000 km along the planet's equator – Ishtar Terra, and Lada Terra; in many ways they represent Venus's version of Earth's continental regions, but have a totally different origin and composition. They show only moderate signs of volcanism and are structurally complex.

The highest mountain chain on Venus, Maxwell Montes, lies in Ishtar Terra and was formed by a combination of compression, expansion, and lateral movement. However, there is no evidence to suggest any form of plate collision was involved in its formation. Tectonism is also revealed by the presence of **ridge belts.** These are elevated well above the mean surface, are hundreds of kilometres wide and thousands of kilometres long. Two major concentrations of these exist: one in Lavinia Planitia near the southern pole, and the second adjacent to Atalanta Planitia near the northern pole.

Magellan also confirmed areas of extensive complex *ridged terrain* (also known as tesserae), which account for 10% of the planet north of latitude 30° N. It is also to be found in Ovda, Thetis, Phoebe, Beta and Asteria Regiones and also Lada Terra. Comprising a complex network of intersecting troughs and ridges, this type of landscape occurs in elevated blocks with steep sides. The ridges are compressional features and appear to represent regions of thicker crust that were preferentially strained in response to regional deformation (Figure 7.4).

Faulting also characterizes certain of the major **volcanic rises**, such as Beta Regio, which measures 2000 km × 3000 km across and

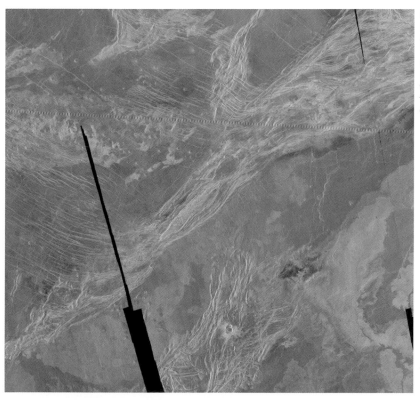

Figure 7.4 Magellan high resolution image of ridge belts in the Lavinia region. The region measures 540 km north to south and 900 km from west to east. PIA37601. Courtesy NASA/JPL.

rises to at least 5 km above datum. Beta, Atla, Eistla and Bell Regiones represent broad topographic swells showing moderate tectonic deformation, shield-type volcanism and deep levels of compensation. The rifting associated with Beta is comparable in dimensions to Earth's East African Rift and gives a relief difference of 6 km in places.

Almost certainly no primeval crust remains on Venus. It has been widely hypothesized that Venus underwent some sort of global resurfacing about 300–500 million years ago, (an age based on crater counting). One possible explanation for this is that it is part of a cyclic process on Venus. Various models have been proposed: a catastrophic model suggests resurfacing occurs infrequently but at extremely high rates; a 'leaky planet' model suggests resurfacing events are global and occur at moderate rates; the 'regional resurfacing' model postulates that regional lows were periodically infilled by volcanic materials of significant volume.

On Earth, plate tectonics allows heat to escape from the mantle. However, Venus lacks such a process, therefore the interior heats up due to the decay of radioactive elements until material in the mantle is hot enough to force its way to the surface. The subsequent resurfacing event covers most or all of the planet with lava, until the mantle is cool enough for the process to start over.

Earth

Earth is the body about which we know the most, and geologists have gradually built a model for how the planet works, a model called **plate tectonics**. The internal structure of our planet is well constrained, largely by seismic studies, and rock samples have established

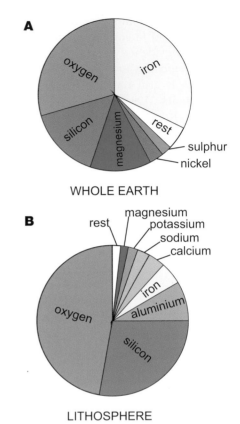

WHOLE EARTH

LITHOSPHERE

Figure 7.5 (a) The chemical composition of the whole Earth. **(b)** The chemical composition of the lithosphere.

its bulk and detailed chemistry (Figure 7.5). Radiometric methods have provided geologists with a technique for dating rocks and building up a history of the planet's geological evolution, while paleomagnetic studies have helped in establishing where continents and oceans were at any point in geological time. None of its primordial crust remains, having been recycled by the processes about to be described. The geological timescale that applies to Earth is shown as Figure 7.6.

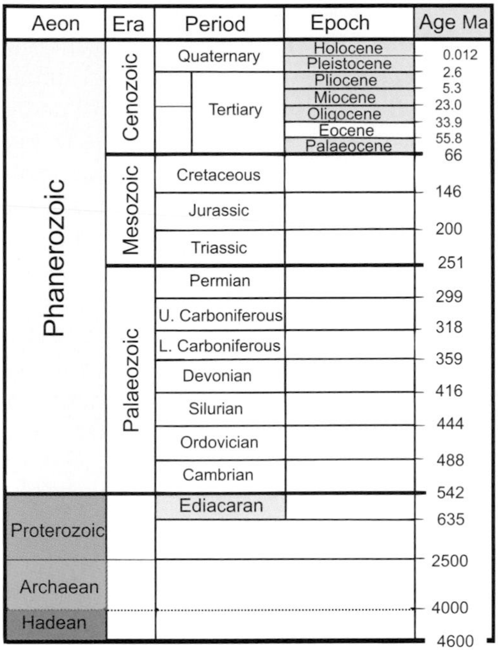

Aeon	Era	Period	Epoch	Age Ma
Phanerozoic	Cenozoic	Quaternary	Holocene	0.012
			Pleistocene	2.6
			Pliocene	5.3
		Tertiary	Miocene	23.0
			Oligocene	33.9
			Eocene	55.8
			Palaeocene	66
	Mesozoic	Cretaceous		146
		Jurassic		200
		Triassic		251
	Palaeozoic	Permian		299
		U. Carboniferous		318
		L. Carboniferous		359
		Devonian		416
		Silurian		444
		Ordovician		488
		Cambrian		542
		Ediacaran		635
Proterozoic				2500
Archaean				4000
Hadean				4600

Figure 7.6 The Earth's geological timescale.

The Earth's oceans and atmosphere are very complex and interact continuously; they also interact with the solid Earth, for there is a constant flow of chemical compounds into the ocean. Much is weathered from solid rocks and dissolved in sea and river waters, much also is in the form of sediments worn from the continental regions and deposited on the sea floor. There it is buried, compressed and converted into rocks that add volume to the **lithosphere**. Later it may be destroyed by the processes that generate plate activity.

The less dense silicate minerals tend to be concentrated in the continental crust, giving it a mean density of 2.7 gm/cm³. It is comprised of the rock granodiorite, containing around 65–69% SiO₂. The denser, more refractory compounds have been concentrated in the oceanic crust, which has a mean density of 2.9 gm/cm³, and composed of the rock basalt with a silica content of between 48 and 50%. Andesite, the magma typical of the continental margins of the Pacific Ocean, is intermediate in composition, with a silica content of between 57 and 63%.

Modern plate theory involves the generation, movement and recycling of Earth's two prominent structural forms, continental and oceanic crust. The theory states that the Earth's outer layers, comprising the crust and uppermost layer of the mantle, together called the lithosphere, slowly move around, driven by convective motions in the underlying mantle layer. It is now established that a series of huge plates make up the outer skin of our planet. At some boundaries these are pulling apart and new sea floor is being created above upwelling convection cells, while at others, ocean floor is being consumed or subducted beneath continental margins, there to be recycled inside the Earth, as plunging cells drag down the outer layers (Figure 7.7).

The thermal gradient –between the Earth's centre and the outside (average temperature increase 2–3°C per 100 m depth) is the major factor in the generation of mantle convection. The Earth's core has a temperature of around 6000°C, while at a depth of around 100 km, in the layer known as the asthenosphere, temperatures of 1100–1200°C prevail. This is the level at which melting begins inside the Earth and from whence most magmas appear to derive.

The convective overturn of the mantle has played a vital part in giving us the Earth we now have. Mantle convection brings thermal energy from depth towards the mantle/

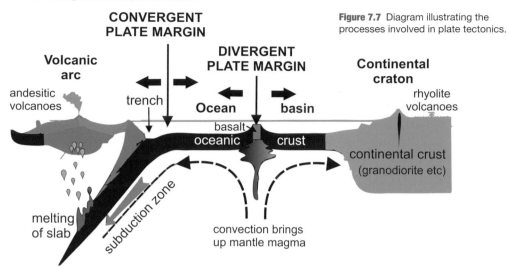

Figure 7.7 Diagram illustrating the processes involved in plate tectonics.

lithosphere boundary. When the rising cells reach the underside of the lithosphere they are forced to spread sideways, cooling as they do so, and applying an immense frictional drag on the underside of the brittle lithosphere. Where two uprising cells diverge, the overlying lithosphere is stretched apart and pressure on the rising mantle material reduced, allowing it to partially melt. This melting generates new oceanic crust along submarine fractures.

The melting of the rising mantle material is brought about by the release in pressure that it experiences when it approaches the surface. This process, termed **pressure release melting**, may occur wherever the outer skin of the planet is breached, allowing the mantle material to partially melt. Partial melts occur because only a fraction of the mantle material will liquidise: very high temperatures would be required to melt it all. So the magma that forms will contain the matter that has the relatively lower melting points. This generates a melt of basaltic composition. Such melts (magmas) form at about 1100°C and will crystallize out minerals such as olivine [$(FeMg)_2SiO_4$], pyroxene [$CaMgFe(Si_2O_6)$] and plagioclase feldspar [$NaCa(Al_2Si_2O_8)$]. These are the essential ingredients of the rock basalt.

In contrast, where the cooled cells start to dive down towards the core/mantle boundary, the lithospheric slabs converge, generating immense pressures that deform the surface rocks trapped between them. This usually involves accumulated sea floor sediments and volcanic deposits that collect in deep **subduction trenches** and may be several kilometres thick. These become deformed and thrust as prismatic bodies onto the continental margins, as **accretionary prisms**. Such structures often include the deposits of massive submarine landslides (Figure 7.8). Similar materials become deformed into new mountain chains when an ocean is completely destroyed as two continental massifs converge.

Figure 7.8 Huge blocks of disrupted quartzite in submarine landslide, Lleyn Peninsula, North Wales.

Four major scientific developments spurred the formulation of what was to become the modern theory of plate tectonics. Firstly, there was new data concerning the bathymetry of the world's ocean floors; secondly, confirmation that there had been repeated reversals of the Earth's magnetic field; thirdly, the development of the hypothesis of **sea-floor spreading** and, finally, realization that earthquakes and volcanism are concentrated along deep oceanic trenches and submarine mountain ranges known as oceanic ridges.

Sea-floor spreading generates new oceanic crust at mid-oceanic ridges, which form a submarine mountain chain that encircles the globe. The ridges are broad and high and usually have central rifts and many volcanic structures, including very vigorous hydrothermal vents called 'black smokers' near their crests. They are regions of very high heat flow.

Both **radiometric dating** of the oceanic crust and the study of paleomagnetism enabled geologists to understand what was happening. It has been shown that for all oceans, the rocks are older the greater the distance from the adjacent oceanic ridges, and it was established that the oldest oceanic crust was a mere 220 Ma old. Subsequently geophysicists began running geomagnetic surveys across different parts of the ocean basins. The first such studies took place both off the coast of California and south of Iceland, on the Reykjanes Ridge, and showed how the strength of the magnetic field changed abruptly but identically on either side of mid-oceanic ridges. The field strength

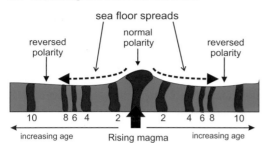

Figure 7.9 The process of sea-floor spreading. Fresh magma rises at the mid-ocean ridge, takes on the magnetization of the ambient field, then spreads sideways due to convection in the underlying mantle. Normal magnetization (red), reversed (blue).

changes were interpreted to show that the Earth's magnetic field had reversed many times in the past. Such **magnetic reversals** were the vital pieces of the jigsaw that put in place the modern theory of sea-floor spreading and oceanic crust generation (Figure 7.9). It confirmed that new oceanic crust emerges at ridge crests and is slowly dragged sideways by the convective mantle motion, at rates of between 4 and 13 cm/yr.

Since the Earth is not growing and new lithosphere is being created, something must be happening elsewhere for the constant size of the Earth to be maintained. Lithospheric destruction occurs where the mantle convective cells plunge downwards, for here the lithospheric plates are being dragged towards one another. Generally one of the converging slabs is made from dense oceanic crust and the other of buoyant continental material; in this case the oceanic plate will plunge down beneath the continental margin along an inclined plane known as a **subduction zone**.

The immense frictional movements along such planes generate major earthquakes that increase in depth towards the continental margin; they also crush and deform the deposits of the ocean floor that get dragged down the subduction zone and, when the subducted slab gets dragged deep enough, the slab will begin to melt, forming new magma. Being more buoyant than the enclosing mantle, this will rise towards the surface, fuelling volcanic activity. Because it has the opportunity to react with the crust through which it rises, magmas generated in such environments tend to be of andesitic, or intermediate, composition.

Not surprisingly, terrestrial volcanism tends to be concentrated along plate boundaries; but there are exceptions. The Hawaiian volcanoes are a case in point. These sit within the Pacific plate, nowhere near a margin. The presence of a **mantle plume** or **hot spot** is invoked to explain this. Other such spots are believed to exist, and may explain the occurrence of some volcanic centres located away from plate boundaries.

Over the eons of time, oceans and continents have grown and shrunk, as the convection motions within the planet have changed. Today some oceans are shrinking (e.g. the Pacific), while others are growing, (e.g. the Atlantic and Red Sea). In the past many regions of oceanic crust have been completely consumed inside the Earth, whereupon adjacent continents have collided, one being accreted onto the other. This occurred when India collided with the southern boundary of Eurasia, generating the mighty Himalayan ranges. Currently as Africa approaches the southern boundary of Eurasia, the Mediterranean crust is being consumed and, roughly 55 Ma from now, may have totally disappeared, whereupon a new mountain range will rise where the Mediterranean now sits. Because collisions constantly add new material to the continental margins, this is why the ages of continents increase inwards; they grow by a process of slow accretion.

Mars

Ongoing spacecraft exploration of the Red Planet bears witness to the fascination of scientists with this rocky world. The possibility that life once developed there has occupied the thoughts of many writers since Schiaparelli first described the (supposed) Martian canals back in the late 1890s. Even today, when the canals have been shown not to exist, the search for signs of life is ongoing.

It was not until the first images were returned by Mariner 4 in 1965 that it was realized the planet had a cratered surface; then, a few years later (1971), Mariner 9 was greeted with the amazing sight of giant volcanoes. Since then there have been many more missions and our knowledge of this small world has increased hugely; currently NASA's roving vehicle, Curiosity, is trundling its way across the surface near Gale crater. It has sampled both soil and rocks and imaged the surrounding terrain.

Mars has an equatorial radius of 3393.4 km but is pear-shaped. A huge bulge characterizes the hemisphere, wherein lies Tharsis, and a similar bulge exists in the region of Elysium. Tharsis rises 10 km above datum and measures 4000 km across. Elysium is smaller. Both show a concentration of major volcanic edifices. The planet has marked topographic asymmetry, the majority of the southern hemisphere lying 1–3 km above datum and most of the northern hemisphere below it. The line of dichotomy separating these two elevation zones describes a great circle inclined 35° to the equator. It is a zone of steep slopes and considerable geological activity. A topographical map of the planet is shown in Figure 7.10.

The southern hemisphere has been intensively cratered and the landscapes that remain have been carved from this ancient surface. It is, however, unlike the lunar highlands, since between the impact craters are channelled plains of various kinds. The

Figure 7.10
Topographic map of Mars derived from the Mars Orbiting Laser Altimeter (MOLA). Courtesy NASA/GSFC.

channel networks clearly were incised by running water. Some intercrater plains show the development of ridges, in appearance very similar to lunar **wrinkle ridges,** and known to be of volcanic origin. These ridged plains cover an area of approximately 29 million km².

The major impact basins of Hellas and Argyre are the exceptions to the positive topography of the southern hemisphere. Hellas is 9 km deep and Argyre 4 km. Several ancient caldera structures also are to be found within this hemisphere, amongst which are Amphitrites, Hadriaca and Tyrrhena Paterae. Evidently there was an early phase of centralized volcanism on Mars.

What might be the age of these intercrater plains? Well, to date no rock samples from Mars have been dated, and all quoted ages are based on impact crater counting. This makes certain assumptions about impactor flux in the distant past, and is often calibrated with lunar data. Martian time has been subdivided into several periods, the oldest of which is the Noachian (*see* table 7.1). The degree to which the quoted dates match reality is unknown and should be treated with caution.

The most significant local variation on the planet is the great equatorial canyon system, Valles Marineris. Plunging to 7 km below datum, this complex of canyons extends for 4500 km along the equator. It is four times deeper, six times wider and at least ten times longer than the Grand Canyon of Arizona. It is estimated that the elastic lithosphere beneath it was less than 30 km thick at the time of its formation. The general physiography is shown in Figure 7.11. Originally the canyon system was believed to have been an erosional feature, having been either eroded by water, or **thermokarst**. However, as imagery improved it became clear that the landscape was generated by rifting, the rift faults subsequently having been modified by collapse and erosion. Huge scalloped embayments bear witness to major wall failure; one massive slide, in Ius Chasma, is 100 km wide and extends right across the floor of the chasm.

It is very clear that the lower northern hemisphere, characterized by plains deposits with few impact craters, and incised by channel networks, has been the subject of major resurfacing. Several ideas have been proposed to

Table 7.1 Ages of Martian plains as derived by Carr (1981).

Plains region	Number of craters <1km/10⁶ km³	Years (billion)	Range
Mare Acidalium	830	1.2	0.2 – 1.7
Sinai Planum	970	1.4	0.4 – 3.0
Utopia Planitia	1270	1.8	0.6 – 2.3
Noachis Planitia	1740	2.5	0.9 – 3.6
Amazonis Planitia	1940	2.8	1.0 – 3.7
Syrtis Major Planum	2053	2.9	1.2 – 3.7
Chryse Planitia	2100	3.0	1.2 – 3.8
Lunae Planum	2400	3.5	1.7 – 3.8
Hellas	2640	3.8	2.9 – 3.9
Hesperia Planum	2710	3.9	3.0 – 3.9

Figure 7.11 Viking Orbiter colour mosaic of Valles Marineris. This is over 4000 km long and 8 km deep. Courtesy NASA/JPL.

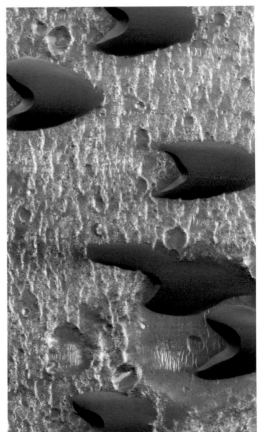

account for this; one suggests an early impact was responsible, another that first order convective overturn caused a 'plate' to subside in the northern hemisphere. Regardless of whatever process produced the lower hemisphere, its original topography has been buried by a very lengthy period of deposition. Some plains are intensively ridged, while others show features that can only be associated with volcanism. That there has been considerable aeolian action is without doubt. Vast dune fields are abundant and often impressively large (Figure 7.12). The plains surrounding the northern pole, which have a muted aspect and lack obvious prominent landforms, clearly are built from a succession of volcanic, aeolian and alluvial sediments, but aided by widespread glaciations. The extent to which ice sheets covered the surface of Mars in times past is an area of robust discussion.

Characteristic of Mars are the planet wide dust storms. It was one of these that was blowing when Mariner 9 arrived at Mars in 1971. These global events tend to coincide

Figure 7.12 Mars Reconnaisance Orbiter HiRise image of barchans dunes near Marwth Vallis. Courtesy NASA/JPL-Caltech/Univ. of Arizona Caption: Alfred McEwen.

with the perihelic retreat of the southern polar cap. Due to the eccentricity of Mars's orbit, insolation is 40% greater at perihelion and leads to enhanced wind activity. Furthermore, near perihelion a large temperature gradient exists between the newly exposed region around the pole and the residual ice cap.

Various studies of the northern plains have described potential strand lines for ancient lakes and oceans. This is a contentious area of research, but the suggestion has been made that there have been at least two periods of previously higher 'sea level', possibly as recently as early Amazonian times. The channels that debouch from the Chryse region, alone appear to have produced sufficient water to fill a substantial body of standing water (assuming the climate was temperate at the time). It has

also been suggested that ancient glaciers once existed. To date no incontrovertible evidence has been produced for either claim, but the possibility still exists.

Water and volatiles on Mars

While water is rare or absent on Mars at the present time, there is irrefutable evidence that in times past large volumes flowed across the planet's surface. Major channel networks emerge from the equatorial regions, becoming wider and shallower downstream. These **outflow channels** are a manifestation of major episodes of flooding. Mars Pathfinder, which landed at the mouth of Ares Vallis, provided unambiguous evidence that fluvial activity had distributed debris along its course (Figure 7.13). Such channels appear

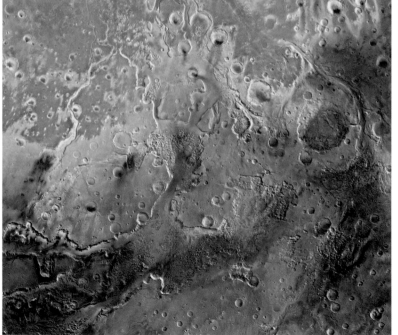

Figure 7.13 Viking Orbiter image of outflow channels in southern Chryse. Valles Marineris lies to the left of the image. PIA00418. Courtesy of NASA/JPL/USGS.

to have been cut over a wide range of ages: some were emplaced into the cratered highlands – predating the end of the major phase of impact cratering – while others debouch onto lightly cratered plains such as those of Amazonis. The impressive channels of the Lunae Planum–Chryse region may well have been developing over much of Martian history, and the suggestion has been made that the highlands peripheral to the Chryse Basin may have acted as a recyclable aquifer via a process of groundwater 'creep'.

Rather different are **fretted channels**. These develop from areas of chaotic terrain close to the dichotomy boundary. They have flat floors and are characterized by extensive debris aprons and flows: evidence for an origin in mass wasting (Figure 7.14). They provide firm evidence for the existence of sedimentary material that has moved under the influence of gravity and probably without the agency of water.

The channel landforms that most closely resemble terrestrial fluvial systems are the smaller **valley networks** that typify the elevated cratered hemisphere (Figure 7.15). The smaller ones are between 1 and 2 km wide but form networks 100 km long or more. Most are concentrated between latitudes 30° N and 40° N and appear to be ancient features. One or two larger networks such as Nirgal and Ma'adim Valles may have started off as run-off channels but have been modified much later by wall retreat.

Recent results from NASA's Curiosity rover, which is exploring the Martian surface near Gale crater in the region of Aeolis Palus, have given information about minerals present at the surface. A rock that was crushed by the rover's wheels exposed a white surface

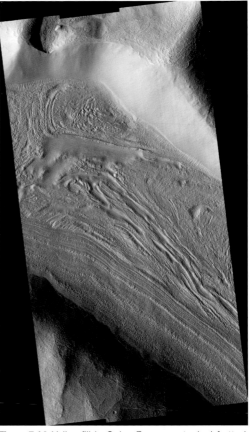

Figure 7.14 Valley fill in Colae Fossae, a typical fretted channel landform. It is probably made of ice-rich material and boulders that are left behind when the ice-rich material sublimates. PIA09367. Courtesy NASA/JPL/Univ. of Arizona.

due to hydrated minerals. Curiosity has also found clay minerals in a drilled sample. Such hydrated silicates indicate formation, or subsequent alteration by, water.

Curiosity's CheMin detector scooped up at a patch of dust and sand that the team named Rocknest. The sample has at least two components: dust distributed globally in dust

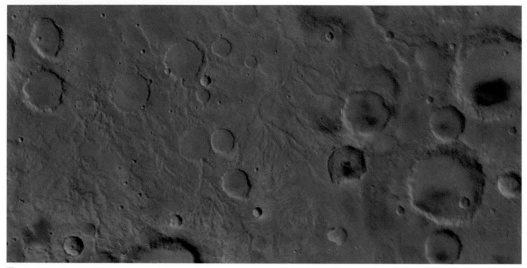

Figure 7.15 Viking Orbiter colour image of channel networks in the southern plains of Mars. PIA00413. Courtesy NASA/JPL/USGS.

storms, and fine sand originating more locally. The dust that was analyzed is mineralogically similar to basaltic material, with significant amounts of feldspar, pyroxene and olivine. Roughly half the soil turned out to be non-crystalline material, such as volcanic glass or products from weathering of the glass. Team scientists felt that, so far, the materials Curiosity has analyzed are consistent with their initial ideas about the deposits in Gale Crater recording a transition through time from a wet to dry environment.

That sedimentary rocks exist on Mars is without doubt. Very recently, observations by the Curiosity rover in Gale crater have revealed isolated outcrops of cemented pebbles (2 to 40 mm in diameter) and sand grains with textures typical of fluvial conglomerates (Figure 7.16). It seems that sediments were transported downhill from the eroding crater rim into a network of streams

that then flowed into a lake environment represented by the mudstone drilled by Curiosity. It is possible that this represents an alluvial fan and may indicate that there once was a body of standing water nearby. NASA scientists suggested it was likely this water ultimately came from melting snow, at a time in the past when significant water-ice may have been present at Mars's equator.

Emission spectra obtained at one outcrop show a predominantly feldspathic composition, consistent with minimal aqueous alteration of sediments. Sediment was mobilized in ancient water flows that likely exceeded the threshold conditions required to transport the pebbles. There is also unequivocal evidence for aeolian deposits, massive debris flows within the canyon system and on the floors of fretted channels. Close to both poles are thick sequences of laminated deposits, and these obscure the older cratered plains

5 cm

Figure 7.16 Curiosity image of the rock Hottah. Evidence for the existence of an ancient stream comes from the size and rounded shape of the gravel in and around the bedrock. The rock has clasts of gravel embedded in it, up to a few centimetres in size and located within a matrix of sand-sized material. Some of the clasts are round in shape, leading the science team to conclude they were transported by a vigorous flow of water. PIA16156. Courtesy of NASA/JPL-Caltech/ MSSS.

of the southern hemisphere down to latitude 80°. These polar plains clearly are a mix of aeolian material, ice and dust. In the southern hemisphere they are thought to be 1–2 km thick, while in the north they are more than double this.

Recent images returned from NASA's Curiosity Rover revealed members of the Yellowknife Bay Formation, and the sites where Curiosity drilled into the lowest-lying member, called Sheepbed. The scene (Figure 7.17) has the Sheepbed mudstone in the foreground and rises up through Gillespie Lake member to the Point Lake outcrop. These rocks record superimposed ancient lake and stream deposits.

Figure 7.17 This mosaic of images from Curiosity's Mast Camera (Mastcam) shows geological members of the Yellowknife Bay formation. The scene has the Sheepbed mudstone in the foreground and rises up through Gillespie Lake member to the Point Lake outcrop. These rocks record superimposed ancient lake and stream deposits. PIA. Courtesy of NASA/JPL-Caltech/MSSS.

Polar layered deposits

The polar ice caps hold a substantial amount of volatile material; much of this is water ice. The southern ice cap, which extends for 1.04 million km², and is thought to be over 1 km thick, has been calculated to hold 2–3 million km³ of ice. The northern cap is less voluminous, but together it is estimated that the total polar volatile inventory is between 3.2 and 4.7 million km³, sufficient to form a global layer of water 22–33 m in depth. Recent research in the region of Deuteronilus Mensae – amidst the fretted terrain – has identified potential glacial deposits emerging from one of the typical amphitheatre valley ends.

The Martian polar regions have layered deposits of ice and dust. The stratigraphy of these deposits is exposed within scarps and trough walls and is thought to have formed due to climate variations in the past. It is thought that the Martian polar layered deposits record climate variations over at least the last 10 to 100 million years.

The north polar layered deposits are layers of dusty ice up to 3 km thick and approximately 840 km in diameter. They are seen exposed in the walls of troughs and scarps (Figure 7.18). It is thought that the north polar layered deposits likely formed recently as rhythmic variations in Mars's orbit changed the distribution of water ice around the planet. As ice moved to and from the polar region in response to a changing climate, layers of ice and dust built up at the poles.

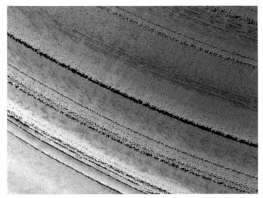

Figure 7.18 The High Resolution Imaging Science Experiment (HiRISE) camera on NASA's Mars Reconnaissance Orbiter recorded this image of north polar layered deposits on March 11, 2010. PIA12997. Courtesy NASA/JPL-Caltech/Univ. of Arizona

The residual ices on top of these layered deposits exert control over the current Martian climate (and vice versa). These ice caps are composed of clean, large-grained water ice at the north pole and carbon dioxide ice at the south. The northern residual ice may be the site of active formation of the polar layered deposits and controls the global distribution of atmospheric water vapour. The southern residual ice cap currently controls the pressure of the atmosphere by supplying extra carbon dioxide gas when it partially sublimates during the summer (Figure 7.19).

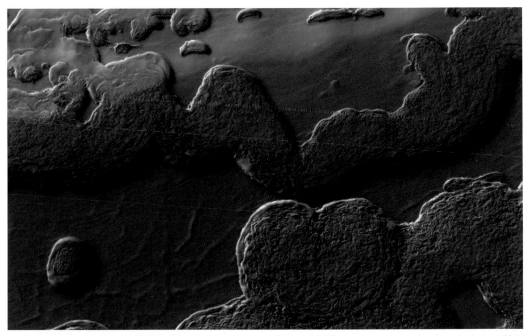

Figure 7.19 This image from NASA's Mars Reconnaissance Orbiter (MRO) shows a variety of surface textures within the south polar residual cap of Mars. It was taken during the southern spring, when the surface was covered by seasonal carbon dioxide frost, so that surface relief is easily seen. PIA13269. Courtesy of NASA/JPL-Caltech/Univ. of Arizona.

8 Volcanism within the inner Solar System

Volcanism is a fundamental process within our planetary system. The injection of molten magma into the upper levels of crusts, and the effusion of silicate magmas has occurred on all of the rocky worlds, including Earth's Moon, and lavas of similar composition to those known from Earth, Venus and Mars are represented in meteorites and asteroids. The volcanic rocks that have crystallized at planetary surfaces and within their crusts represent partial melts of solid mantle materials. These have risen to high levels when the brittle outer layers have been fractured or weakened by internal forces such as mantle convection, or thermal activity.

The extremely widespread occurrence of basalt on all of the terrestrial planets indicates that basaltic partial melts developed from **pyrolite**-type source rocks (table 8.1).

Table 8.1 Chemical composition of pyrolitic mantle source rock.

Oxide	Weight %
SiO_2	38%
Al_2O_3	2%
FeO	5%
MgO	50%
CaO	3%

The implication is that similar source rocks must have been common at upper mantle levels on all of the inner planets, which further suggests a degree of homogeneity in the primordial matter from which they evolved. However, the subsequent evolution of these melts may have followed quite different paths on different planets, due to the very different chemical and physical environments in which they crystallized.

Magmas which rise rapidly from their source regions, to be erupted at the surface as lavas, may chill sufficiently rapidly for their pristine chemistry to be preserved. In nature, most magmas do not behave this way, and exhibit varying degrees of chemical fractionation as they rise towards the surface. Often a volume of melt will stall on its way to the surface and remain in a holding reservoir for some period, during which fractional crystallization may occur. In time the fractionated melt may continue to the surface, having been modified, there to erupt as a quite different magma from its source. In general, fractionated melts contain more silica than their source magmas, as the early-formed, more refractory minerals accumulate at the base of crystallizing magma chambers, having sunk there under the influence of gravity (Figure 8.1).

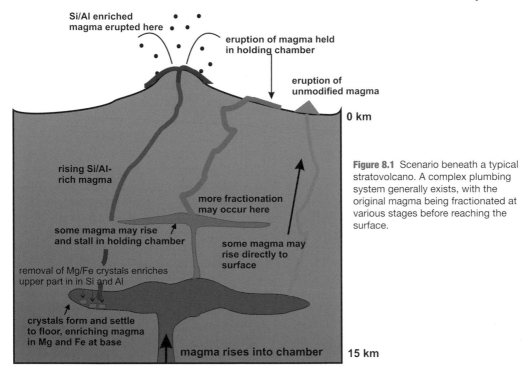

Si/Al enriched magma erupted here

eruption of magma held in holding chamber

eruption of unmodified magma

0 km

rising Si/Al-rich magma

more fractionation may occur here

some magma may rise and stall in holding chamber

some magma may rise directly to surface

removal of Mg/Fe crystals enriches upper part in in Si and Al

crystals form and settle to floor, enriching magma in Mg and Fe at base

magma rises into chamber

15 km

Figure 8.1 Scenario beneath a typical stratovolcano. A complex plumbing system generally exists, with the original magma being fractionated at various stages before reaching the surface.

Mercurian volcanic plains

Spacecraft have revealed there to be two geologically distinct plains units on Mercury.

Intercrater plains are the oldest visible surface, predating the heavily cratered terrain. They are gently rolling or hilly and occur in the regions between larger craters. Such plains appear to have obliterated many earlier craters, and show a general paucity of smaller craters below about 30 km in diameter. They are distributed roughly uniformly over the entire surface of the planet. Furthermore, in some areas they do show evidence for a volcanic origin. Their general appearance suggests that they represent **flood lavas** of basic composition, e.g. basalts or komatiites.

Smooth plains are widespread flat areas resembling the **lunar maria**, which fill depressions of various sizes. Notably, they fill a wide ring surrounding the Caloris Basin. An appreciable difference to the lunar maria is that the smooth plains of Mercury have the same albedo as the older intercrater plains. Despite a lack of unequivocally volcanic features, their localization and lobate-shaped units strongly support a volcanic origin. They formed significantly later than the Caloris basin, as evidenced by appreciably smaller crater densities than on the Caloris ejecta blanket.

The recent NASA MESSENGER spacecraft, which started its close orbital survey in March 2011, mapped a huge expanse of volcanic plains surrounding the north polar region of Mercury. These cover more than 6% of the total surface of Mercury. The volcanic deposits are thick, as is shown by analysis of the size of buried 'ghost' craters. This indicates that the lavas are locally as much as 2 km thick. The plains show the characteristics typical of flood lavas, which appear to have poured out from long, linear vents and covered the surrounding areas, flooding them to great depths and burying their source vents (Figure 8.2).

Vents, measuring up to 25 km in length, appear to be the source of some of the tremendous volumes of very hot lava – possibly komatiite - that must have rushed out over the surface of Mercury and eroded the substrate, carving valleys and creating teardrop-shaped ridges in the underlying terrain. One or two images reveal pyroclastic deposits (Figure 8.3).

Figure 8.2 Mercury MESSENGER image of a small region of the northern plains of Mercury. Here there are many volcanically flooded craters such as Monteverdi, which occupies part of the lower edge of this image. The plains have abundant wrinkle ridges and fresh craters. PIA 17389. Courtesy of NASA/Johns Hopkins University Applied Physics Laboratory/Carnegie Institution of Washington.

Figure 8.3 Mercury MESSENGER image of a bright pyroclastic deposit that is thought to have erupted from a vent near Rachmaninoff crater. PIA 17501. NASA/Johns Hopkins University Applied Physics Laboratory/Carnegie Institution of Washington.

Volcanic plains on Venus

Volcanism has been the dominant process responsible for the plains of Venus and that means 85% of the planet's surface. The few rocks that have been analyzed were tholeiitic basalts, some a little more alkaline than others. Analyses from the Russian Venera and Vega missions are shown in Table 8.2, together with an analysis of a typical terrestrial oceanic basalt.

The plains have different radar signatures, some being brighter than others; this is a reflection of differing degrees of surface roughness rather than any compositional variation, for they generally share similar features, such as long flow-like structures, caldera depressions, low shields and domes. Such landforms are typical of plains-style volcanism as developed on Earth in such regions as the Columbia and Snake Rivers. Their emplacement involved rapid effusion of high-volume basalts or even komatiites in response to extensional tectonics. On Venus compression has modified plains units in some regions.

Over 50 lava channel systems have been imaged, some have lengths >250 km. While many Martian long runout flows have central channels with raised levees, such levees are not found on Venus. Since the width of Venus flows is at least as great or greater than those on Mars, this clearly is not a resolution issue. More likely it is a manifestation of the different rheology of flows on the two planets; the much higher ambient temperature on Venus must have a major part to play in this.

By terrestrial standards, many individual flow fields are huge: volumes as great as 8000 km^3 are not uncommon. One group of flows, south of Ozza Mons, covers an area of 180 000 km^2, with the longest flows measuring 1000 km. Particularly spectacular are the lobate flows in the region of Sapas Mons (Figure 8.4).

Superposed on the extensive volcanic plains are a large number of small shields, domes and cones. The majority of the small shields fall into the size range 2–8 km and are <200 m high. Cones generally have diameters <15 km, may have summit pits, and show rougher surfaces than shields. This suggests they may be composed of scoriaceous lavas. Domes often have scalloped flanks, suggestive of dome collapse; however, the volume of outflow material is an order of magnitude greater than anything found on Earth, and is another reflection of the very different surface conditions on Venus.

Table 8.2 Chemical analyses of Venus surface units.

Constituent	Venera 13	Venera 14	Vega-2	Oceanic basalt
SiO_2	45±3	49±4	45.6±3.2	51.4
Al_2O_3	16±4	18±4	16.0±1.8	16.5
MgO	10±6	8±4	7.7±3.7	7.56
FeO	9±3	9±2	11.5±3.7	12.24
CaO	7±1.5	10±1.5	11.5±3.7	9.4
K_2O	4±0.8	0.2±0.1	0.1±0.1	1.0
TiO_2	1.5±0.6	1.2±0.4	0.2±0.1	1.5
MnO	0.2±0.1	0.16±0.08	–	0.26

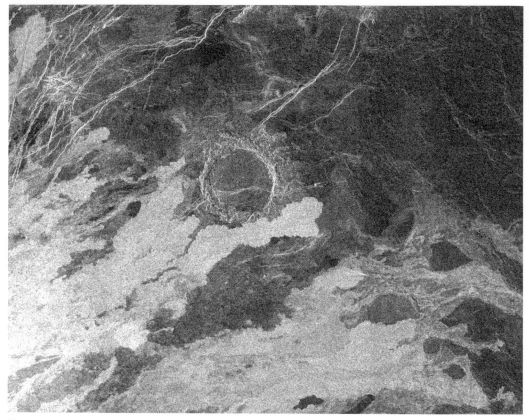

Figure 8.4 Magellan radar image of part of the eastern flank of the volcano Sapas Mons. The bright lobate features along the left-hand side are lava flows that are rough at the 12.6 cm wavelength of the radar. These flows range in width from 5 to 25 km, with lengths of 50 to 100 km. PIA 00099. Courtesy Nasa/JPL.

Centralized volcanism on Venus

The very high atmospheric pressure that prevails on Venus would retard the coalescence of gas bubbles in magmas, so it is very unlikely, from theoretical considerations alone, that explosive volcanism will have occurred. Thus it is that the centralized structures that are found on its surface – domes, vents, shields, coronae, novae, arachnoids and large volcanic rises – appear to have been largely generated by effusive volcanism. The steep-sided nature of some structures, however, may indicate variations of magma composition and viscosity that give rise to such landforms (Figure 8.5).

There are several zones of major rift faulting on Venus, and in some cases, where sets of these intersect, major volcanic rises have developed, e.g. Eistla, Atla and Beta Regiones. These are large and complex volcanic centres

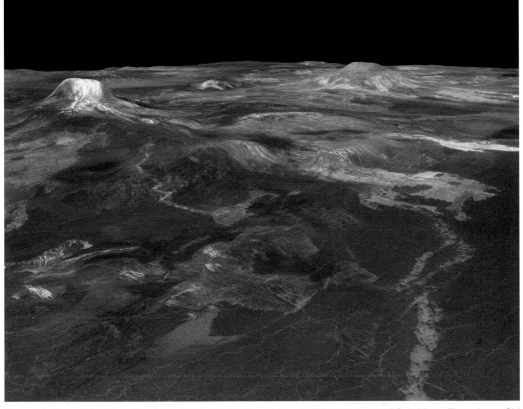

Figure 8.5 Perspective Magellan view of Eistla Regio, viewed from 1100 km northeast of Gula Mons. The volcano Sid Mons can be seen to the right of Gula Mons. P38724. Courtesy NASA/JPL.

and are uniquely Venusian. There is general agreement amongst geologists that these are marked by deep depths of compensation and have their origin in mantle plumes. The term Large Igneous Province (LIP) has been coined to describe them.

In fact, volcanic rises are divisible into three groups based on surface morphology: (i) rift-dominated (Beta, Atla), (ii) volcano-dominated (Dione, Western Eistla, Bell, Imdr), which lack signs of major rifting, and (iii)

coronae-dominated (Central Eistla, Eastern Eistla, Themis), with which several larger than average coronae are associated. Rift-dominated rises show the highest swells (2.5 and 2.1 km) whereas coronae-dominated rises generally show the lowest (1–1.5 km). Beta Regio is a broad rift-dominated volcanic rise and an area of major extensive volcanism located at the intersection of Devana Chasma, a major rift up to 120 km wide and 2 km deep, and a series of W/E older faults (Figure 8.6).

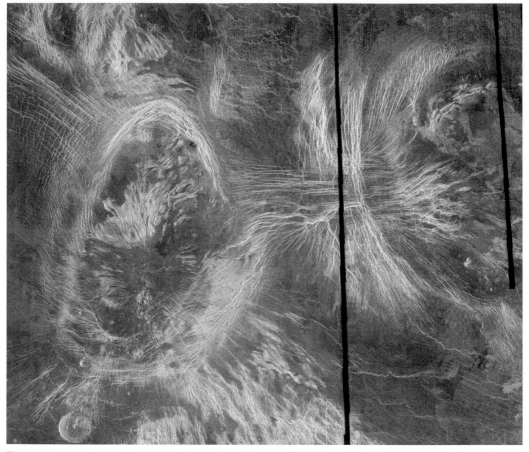

Figure 8.6 Magellan image of Bahet Corona (left), which is about 230 km long, 150 km across, and part of Onatah Corona, which is over 350 km in diameter. PIA 00461. Courtesy NASA/JPL.

The massive volcanic shield of Theia Mons is superimposed on the rift. Associated with this are extensive lobate flows, indicative of widespread basaltic volcanism.

There are also many large volcanic structures between 50 and 300 km across, most being less than 1 km high. These massive shield volcanoes are equivalent to the shields found on Mars and to those of Hawaii. So far, 156 large volcanoes (100–350 km diameter), 274 intermediate volcanoes (20–100 km in diameter) and 86 calderas have been identified. There are also 515 coronae – major circular structures with annular fractures that range in diameter from 60–2600 km that are considered by many to be located above rising mantle diapirs. It appears that radial fracturing was the first stage in the development of coronae, a response to crustal uplift above rising mantle plumes. Subsequently, volcanism developed around such centres, while at a later stage lava flooding sometimes buried most of the radial fractures (Figure 8.7).

The 259 **arachnoids** – circular structures surrounded by concentric fractures (which may include dyke swarms) beyond which is strong radial fracturing – and 64 **novae** – structures that lack annular faults but have a central dome with radial fractures – may represent an evolutionary sequence with the coronae, all being generated in response to mantle upwelling and the interaction of mantle **diapirs** with the base of Venus's lithosphere.

There are also many steep-sided pancake domes. These flat-topped landforms may have a summit depression. Over 150 have so far been identified; they range in diameter between 10 and 70 km and have a mean height of 700 m. This means that the larger of these edifices have volumes >100 km³, an order of magnitude greater than even the most voluminous terrestrial eruptions, such as the Laki fissure in Iceland. The domes tend to be concentrated in regions where there are coronae, the implication being that they too are in some way related to regions of mantle upwelling. Their large dimensions and morphology suggest that they were not formed during a single magmatic event, but many.

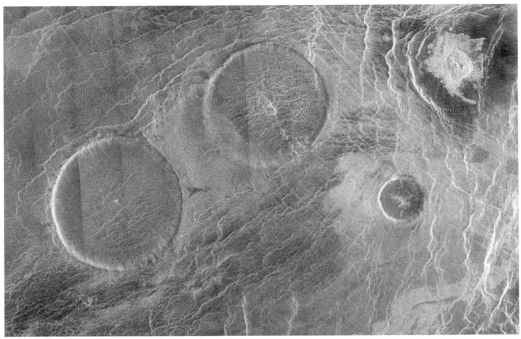

Figure 8.7 Magellan image of large pancake domes about 65 km across in Eistla Regio. These rise 1 km high and are presumed to have formed by extrusion of rather viscous lava. The area shown measures 160 x 250 km. PIA00084. Courtesy NASA/JPL.

This perhaps provides evidence for the fact that some measure of magma fractionation has occurred on Venus, generating melts more evolved and therefore more viscous than basalt.

In this context it is interesting that ESA's Venus Express spacecraft recently studied the geology of the rugged highland terrain called Chimon-mana Tessera and its surrounding volcanic plains. Using near-infrared observations collected by the Venus Monitoring Camera (VMC), scientists have found evidence that the planet's rugged highlands are scattered with rocks of higher silication than the basaltic rocks typical of the volcanic plains.

Volcanism on Mars

The global distribution of volcanoes on Mars is very different from that on Earth. The segmented nature of the terrestrial lithosphere dictates that volcanism tends to be concentrated along linear zones. Since plate activity is absent on Mars, volcanism is probably related to long-lived mantle plumes, and such linearity is not seen.

The earliest recognizable volcanic plains were emplaced in later Noachian times and were succeeded by the ridged plains of Lower Hesperian age (Figure 8.8). Both represent large-volume flood lava eruptions in the southern hemisphere. At a similar time, centralized volcanicity generated the ancient

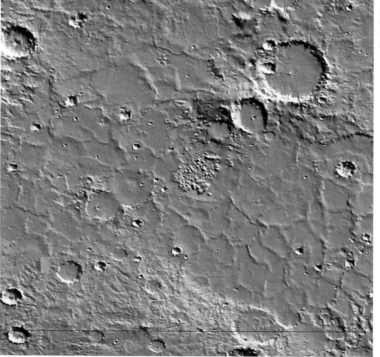

Figure 8.8 Hesperian ridged plains units in the southern hemisphere of Mars. Courtesy NASA/JPL-Caltech/Arizona State University.

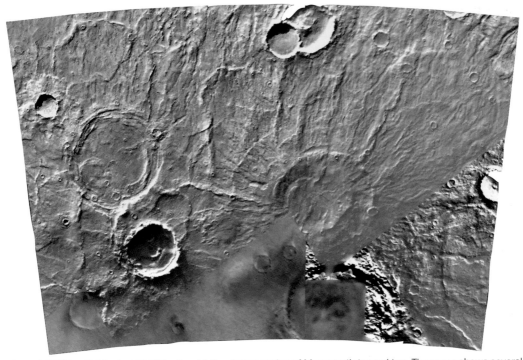

Figure 8.9 Viking Orbiter image of the Amphitrites Patera region of Mars; north toward top. The scene shows several indistinct ring structures and radial ridges of an old volcano named Amphitrites Patera. PIA00410. Courtesy NASA/JPL/USGS.

volcanoes of Amphitrites Patera and Hadriaca Patera, followed by what are believed to be mixed ash and lava volcanoes around the Hellas Basin (Figure 8.9). This early phase of central volcanism appears to have been largely hydromagmatic, implying that conditions on the planet were very different then from what they are now.

Towards the end of Hesperian time, a further phase of widespread flood volcanism laid down the volcanic plains of the northern hemisphere. This was followed by the growth of major central volcanoes in the regions of Elysium (Albor Tholus) and in northern Tharsis (Alba Patera). While a measure of

explosive activity continued, there was a decline in this style towards effusive activity, with extrusion of vast volumes of basaltic lava along the crest and margins of the Tharsis rise, in Elysium, and finally at Olympus Mons, the tallest volcano in the Solar System.

Major volcanic shields are crowned by complex caldera depressions, a manifestation of successive periods of eruption followed by summit collapse (Figure 8.10). Flank slopes generally are low (2–5°) and individual flows often have either levees or axial tubes or channels. Some flows, for instance those that emanate from Alba Patera, extend over 1000 km from its summit. The extreme

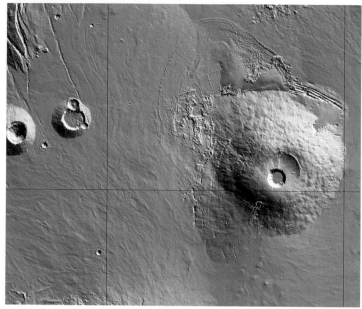

Figure 8.10 The shield volcano, Pavonis Mons (right) has a summit caldera 45 km across and 4.5 km deep. To the west are the smaller shields of Biblis Patera (extreme left) and Ulysses Patera. Courtesy NASA/JPL-Caltech/Arizona State University.

volume of these flows – some tube-fed flows have volumes of 3500 km³ – implies very low viscosity and may indicate komatiite compositions.

The lower gravity of Mars generates lower buoyancy forces on magma rising through the Martian crust, and therefore the magma chambers that feed volcanoes on Mars are thought to be deeper and much larger than those on Earth. If a magma body on Mars is to reach close enough to the surface to erupt before solidifying, it must be big. Consequently, eruptions on Mars are less frequent than on Earth, but are of enormous scale and eruptive rate when they do occur. Somewhat paradoxically, the lower gravity of Mars also allows for longer and more widespread lava flows. Lava eruptions on Mars may be several orders of magnitude greater than those on Earth.

It is likely that the earliest volcanoes, i.e. those of Hellas, were controlled by deep-seated faulting produced during the excavation of the Hellas impact basin. However, volcanism around Tharsis and Elysium undoubtedly is related to the growth of major crustal upwarps, the formation of which still remains somewhat contentious. However, what is clear is that the immense volcanic loads that built up in regions such as Tharsis led to crustal flexure and sometimes failure. Annular faulting typifies the surface of the lithosphere just beyond a volcano's periphery, whereas radial fracturing develops near the base of the elastic lithosphere beneath the volcano's core. Surface collapse is also a feature of some of the larger structures, while the spectacular blocky aureole that surrounds Olympus Mons may represent a major gravity slide.

Five years of Mars Express gravity mapping data are providing unique insights into what lies beneath the Red Planet's largest volcanoes. The results show that the lava grew denser over time, and that the thickness of the planet's rigid outer layers varies across the Tharsis region. The measurements were made while Mars Express was at altitudes of between 275 and 330 kilometres above the Tharsis volcanic 'bulge' during the closest points of its eccentric orbit, and were combined with data from NASA's Mars Reconnaissance Orbiter. The Tharsis bulge includes Olympus Mons – the tallest volcano in the Solar System, at 21 km – and the three smaller Tharsis Montes that are evenly spaced in a row (Figure 8.11). The region is thought by some scientists to have been volcanically active until relatively recent times.

Overall, the high density of the volcanoes corresponds to a basaltic composition that is in agreement with the many Martian

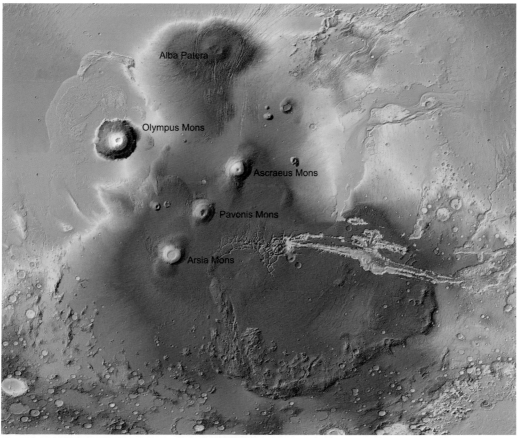

Figure 8.11 MOLA image of the region of the Tharsis Bulge. Recent gravity analysis by Mars Express suggests the three main shields may have been emplaced from SW to NE. Courtesy NASA/MOLA Science team.

meteorites that have fallen to Earth. The new data also reveal how the lava density changed during the construction of the three Tharsis Montes volcanoes. They started with a less dense, possibly andesitic, lava and were then overlaid with heavier basaltic lava that makes up the visible surface of the Martian crust.

The view from the Curiosity rover towards Mount Sharp evidently is over weathered terrain that has broken down in situ to fine-grained mineral clasts but is intermixed with wind-blown dust from further afield (Figure 8.12).

Figure 8.12 Curiosity image of sandy ground en route for Mount Sharp. Image PIA17362. Courtesy NASA/JPL.

Planetary comparisons

The four inner planets all show evidence for extensive volcanism. Mercury's surface is largely built from volcanic plains, punctuated by impact basins and craters. These plains show ridging and some evidence of localized central volcanism. However, major volcanic constructs are absent and plume-related processes appear not to have operated. Venus, in contrast, shows widespread evidence for volcanic plains, but also major volcanic rises and a plethora of major central volcanoes, several kinds of which are uniquely Venusian. Both tectonism and mantle plume-related activity appear to have played a part in volcanism on the planet. Furthermore, there is strong evidence that large areas of the surface have been resurfaced, possibly more than once, the latest episode of which may be geologically recent.

Mars, with its greater inventory of volatiles, has been volcanically active over a lengthy period. Early plains development by flood-style basic lavas predated a move towards centralized activity. Following the resurfacing of the northern hemisphere by outpouring of flood basalts, the massive shield volcanoes of Tharsis and Elysium grew, giving rise to some of the most spectacular landforms within the Solar System. The large volumes of their associated flows suggests that volcano growth occurred above mantle-related plumes that remained in place over lengthy periods.

Earth differs from the other three rocky worlds due to its development of plate tectonics, whereby oceanic and continental crust is constantly recycled. Plume-related volcanism is not the major source of volcanic landforms. Instead, the emergence of mantle-derived magma from divergent plate boundaries resurfaces large regions of its lithosphere, giving rise to new oceanic crust. Where plates converge, oceanic crust and associated sediments are subducted back inside the Earth, partially melted, and rise diapirically along linear zones of andesitic volcanoes, most spectacularly around the Pacific Ocean – the 'Ring of Fire'. Subduction-related volcanism also occurs as island arcs, such as those of Indonesia or the Aeolian Islands.

The presence of abundant water on Earth has played a vital part in the plate recycling process that generates the styles of volcanism seen there. Venus, lacking this volatile, never developed plate tectonics; neither did Mercury. Mars, a planet with a significant inventory of volatiles, including water, also appears to be a one-plate planet; however, some scientists have suggested that first-order convection may have resurfaced the northern hemisphere.

9 Earth's Moon

Earth's one satellite is worthy of a separate chapter because so much has been learned about both it and the Earth from the intensive studies that were carried out during the earlier Apollo and later Clementine mission. It is also the only other world for which we have radiometric dates and detailed mineralogical and chemical analyses of its rocks.

The origin of the Moon

Until the 1980s it was generally believed that the Moon was captured by the Earth at an early stage in its evolution. One problem with this idea is understanding the capture mechanism: a close encounter with Earth should result in either collision or altered trajectories. Furthermore, this hypothesis has difficulty explaining the essentially identical oxygen isotope ratios of the two worlds.

Then, in 1984, a new hypothesis exploded into popularity at a conference held in Hawaii, although two key papers on the topic had been published a decade earlier. The basic premiss of this was that the early Earth was impacted by a large planetesimal that generated a ring of debris that eventually settled into the Earth–Moon doublet. In this way it is possible to account for the amount of angular momentum in the Earth–Moon system, and explain the small size of the lunar core (Figure 9.1).

To account for the amount of angular momentum in the Earth–Moon system, it is

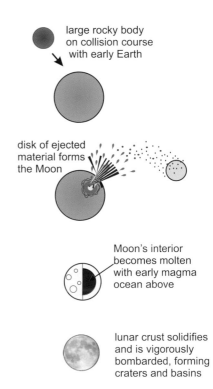

large rocky body on collision course with early Earth

disk of ejected material forms the Moon

Moon's interior becomes molten with early magma ocean above

lunar crust solidifies and is vigorously bombarded, forming craters and basins

Figure 9.1 Collision theory for the origin of Earth's moon.

estimated that the object would need to be about 10% the mass of Earth, that is, about the size of Mars. One of the most important predictions of the giant impact hypothesis is that the Earth was mostly molten when it formed. This would have led to complete separation of elements that concentrated into metallic iron when the core formed.

The Moon's relatively small iron core is explained by the object's core accreting onto Earth's. The lack of volatiles in the lunar samples is also explained in part by the energy of the collision. The energy liberated during the re-accretion of material in orbit about the Earth would have been sufficient to melt a large portion of the Moon, leading to the generation of a magma ocean. The newly formed moon is calculated to have orbited at about one-tenth the distance that it does today, and became tidally locked with the Earth, where one side continually faces toward the Earth.

Recently, this hypothesis has been called into question by measurements that find that the Earth and Moon share the same isotopic composition. The isotopes of oxygen and titanium, for example, vary widely in the Solar System and are used to 'fingerprint' different planets and meteorite groups. The data show that the Earth and Moon are isotopic twins, but the original giant impact model predicted that most of the Moon was made from the body that struck Earth, which should have had a different isotopic fingerprint. The giant impact model therefore has a major problem. Many modifications to this style of theory have been published but, to date, no one theory appears to explain all the characteristics of the system.

Orbital relationships between Earth and Moon

The Moon has a nearly circular orbit (ellipticity = 0.05) which is tilted about 5° to the plane of the Earth's orbit. Its average distance from the Earth is 384,400 km. The combination of the Moon's size and its distance from the Earth causes the Moon to appear the same size in the sky as the Sun, which is one reason we can have total solar eclipses.

It takes the Moon 27.322 days to complete one orbit around the Earth, and because of the effect on the Moon of terrestrial tidal forces, the same side of the moon always faces the Earth. In other words, it takes the Moon the same amount of time to rotate once as it does for the Moon to go around the Earth once. Therefore, Earth-bound observers can never see the 'farside' of the Moon. The farside had not been seen until orbiting spacecraft were sent there in 1959; Luna-3 transmitted 17 very grainy images back to Earth. In fact, because of libration effects, a small area of the farside limb can be seen at times.

The structure of the Moon

The moon appears to have a very small core that accounts for just 1 to 2% of its mass and has a diameter of 680 km (Figure 9.2). It probably consists mostly of iron, but may also contain large amounts of sulphur and other elements. Surrounding this core is a rocky mantle about 1,330 km thick and made up of dense rocks rich in iron and magnesium, i.e. akin to peridotite. The brittle outer crust averages 70 km thick and is largely brecciated

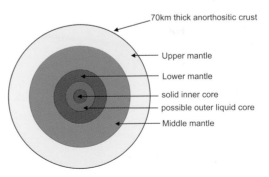

70km thick anorthositic crust
Upper mantle
Lower mantle
solid inner core
possible outer liquid core
Middle mantle

Figure 9.2 The internal structure of Earth's Moon. The inner core is very small and made of iron, nickel and sulphur.

by large impacts; this gives way to more compact material at a depth of 9.6 km. The larger impacts produced many large impact basins, into which rose mantle-derived basaltic magmas, forming the lunar maria. Highlands and maria are the two lunar crustal types.

The nature and origin of the lunar highlands

Even a casual look at the Moon will reveal that the lighter-coloured areas are more rugged than the darker maria. The rugged topography is a direct result of intensive impact cratering early on in lunar history, generating huge numbers of craters and a number of immense impact basins. Much of the geological history of the Moon is controlled by the distribution of large impact structures and their ejecta blankets. Features larger than 300 km in diameter are termed **impact basins**. More than 40 such basins have been identified.

The best-preserved of the large basins is the farside Orientale Basin, which is only partially visible from Earth. Lunar Orbiter images provided our first good look at this basin, which is 930 km in diameter. It is the freshest impact basin on the Moon and is believed to be slightly younger than the Imbrium Basin. It has three concentric rings which are beautifully preserved. Only the innermost ring has been flooded by mare lavas (Figure 9.3).

Figure 9.3 Lunar Orbiter 4 image of the farside basin Mare Orientale. Courtesy LPI/USRA.

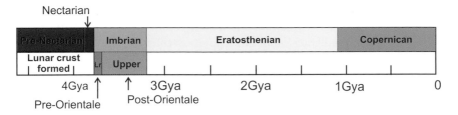

Figure 9.4 A lunar stratigraphy. This is based on the superposition of basin ejecta and crater statistics.

The best-known nearside basin is the Imbrium Basin, which measures 1160 km across and is a prominent feature of the nearside. Imbrium has three concentric rings but has been completely flooded by lavas. It is calculated to be 5.90 km deep with a 4.7 km thickness of mare lava flooding it. Orientale is calculated to be 5.71 km deep with a 0.56 km thick lava fill.

Impact basins also throw out huge volumes of ejecta, and successive blankets of this material can be used to provide a lunar statigraphy (Figure 9.4). The Imbrium basin is particularly useful in this respect, as it is relatively young as basins go, and it has been possible to map the ejecta blanket's extent, at least in those regions not flooded by mare lavas. This basin is believed to have formed 3.85 mya, based on *Apollo 15* data.

Material ejected by the formation of an impact basin is generally redeposited in the region outside the basin. Near the basin rim, the ejecta blanket can be quite thick and completely cover the pre-existing terrain. In the case of Mare Orientale, the thickness of ejecta at the Cordillera Mountains (the rim of the basin) is 2.9 km; further out, at a distance of 215 km, the thickness has dwindled to 1 km. Even in these distal positions, the ejecta can substantially modify the terrain (Figure 9.5).

Figure 9.5 Apollo image of the hummocky ejecta from the Imbrian Basin impact. It is mapped as the Fra Mauro Formation, with an age of 4.25 Ga. Courtesy LPI/USRA.

While *Apollo 11* collected typical mare basalt samples, it also provided the first clues as to the nature of the lunar highlands, for at Tranquillity base were found particles of anorthositic rocks that could only have emanated from the nearby highlands. In addition a third type of rock given the acronym KREEP – standing for potassium (K), rare earth elements (REE) and phosphorous (P) – was found.

Chemistry of highland rocks

The early Moon evolved a magma ocean which solidified into a crust of the lightest minerals which had floated to the top, predominantly aluminium calcium silicates, giving rise upon crystallization to the rock **anorthosite**. This occurred about 4.5 bya. However, the period between 4.5 and 4.0 bya was marked by heavy bombardment by meteors and asteroids, causing intense cratering. Thus the highlands are mostly composed of overlapping layers of material ejected from craters. Because of the intense bombardment, most highland rocks are in the form of breccias, where fragments of different rocks are compacted and welded together by impacts. Rocks brought back from the highlands vary in age between 3.84 and 4.48 Ga.

Chemical analysis of highland rocks reveals them to be rich in refractory elements such as calcium (Ca), Aluminium (Al), and Titanium (Ti) that form compounds having high melting points. However, although anorthosite and anorthositic gabbro are apparently primary rock types, a range of rocks were collected from the highland regions (Table 9.1).

Highland breccias usually show signs of intense shock. Textures may consist of grains that have become so closely packed together as to resemble terrestrial granulites. In other cases very complex fabrics develop, for instance where breccia clasts sit within larger clasts, revealing several phases of brecciation (Figure 9.6). The most extreme effects of impact may produce impact melts.

Table 9.1 Major element contents of typical highland rocks (as oxide %).

Elementc	Anorthosite	gabbro	dunite	troctolite	norite	KREEP	monzonite	granite
SiO_2	44.1	45.3	40.7	42.9	49.8	48.0	59.6	73.1
Al_2O_3	35.5	28.7	1.3	20.7	18.4	14.9	20.6	12.4
CaO	19.7	16.2	1.1	11.4	10.5	7.4	7.9	1.3
FeO	0.2	4.1	11.9	5.0	6.0	9.2	3.3	3.5
MgO	0.1	4.4	45.4	19.1	14.5	7.4	3.3	0.1
TiO_2	0.02	0.3	0.03	0.05	0.08	2.2	0.6	0.5
Cr_2O_3	0.01	0.10	0.11	0.3	0.3	0.3	0.3	0.4
Na_2O	0.3	0.5	0.01	0.2	0.3	0.9	0.9	0.6
K_2O	0.02	0.09	0.002	0.03	0.05	0.6	4.7	6.0

Figure 9.6 Photomicrograph of a typical highland breccia. Left: in plane polarized light and, right, cross-polarized light.

The lunar maria and volcanic activity on the Moon

Prior to the Apollo programme it had been predicted that the low-albedo lunar maria were floored by basaltic lava. However, it was not until *Apollo 11* returned samples to Earth that this was proved to be the case. Later, when the astronauts visited Hadley Rille, layers of basalt were found in the walls of the rille at very shallow depth beneath the lunar **regolith**. The latter is a ubiquitous layer of fine-grained material that ranges in thickness from 5 m on mare surfaces to 10 m on the highlands. The bulk of this is a fine gray soil with a density of about 1.5 g/cm^3, but the regolith also includes breccias and rock fragments from the local bedrock.

Lunar mare materials are exposed over an area of 6.3 x 10^6 km^2 and have an asymmetry such that the greater proportion of the post-Imbrium lavas outcrop on the lunar nearside. Mare lavas occur in a number of situations: (i) they fill or partially fill the central regions of multi-ringed impact basins; (ii) they fill irregular depressions; or (iii) they floor large impact craters. Evidently they were emplaced hydrostatically; they fill depressions. Apollo data indicates that mare volcanism occurred over a period that extended from 3.8 bya to 2.5 bya. Lava thickness varies between 5.2 and 0.63 km, based on Clementine altimetric data.

The lunar regolith

The Moon's surface is covered in a layer of fine material that, at maximum, is a few metres thick. This is termed the regolith, and it formed over hundreds of millions of years by micrometeorite bombardment. It consists of fine rock and glass fragments, mineral clasts and agglutinates. The average grain size is seldom more than a few tens of microns. In some locations, volcanic **fire fountaining** has produced small coloured glass beads such as the orange soil found at Shorty Crater in the Taurus-Littrow valley by *Apollo 17*, and the green glass found at Hadley-Apennine by *Apollo 15* (Figure 9. 7).

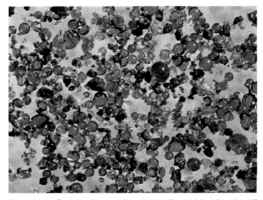

Figure 9.7 Green glass spherules collected by Apollo 17 from Taurus-Littrow. Courtesy NASA/JSC-LRL.

Laboratory analyses of the pyroclastic glasses and beads indicate that, unlike most lunar materials, they have volatile-element enriched coatings from their gas-rich source regions, they originated at great depths (<400 km) inside the Moon, and they represent the most basic or primitive of lunar volcanic materials.

Chemistry of mare basalts

During 1990, the Galileo spacecraft imaged parts of the Moon's western nearside limb and the farside. The solid state imaging experiment (SSI) confirmed earlier data in that most mare lavas are of intermediate-TiO$_2$ composition. Most are rich in mafic constituents, and the two essential minerals are Ca-plagioclase and clinopyroxene. Their textures indicate quite rapid cooling from dry melts, something which sets them apart from typical terrestrial ocean-floor basalts.

Chemically, mare basalts can be divided into two broad groups; the older *high-Ti group* with TiO₂ content 9 to 13% and ages between 3.55 and 3.85 Ga, and the younger, *low-Ti group* (age 3.15 to 3.45 Ga),with TiO₂ contents between 1 and 5%. Samples from the *Apollo 11* and *17* missions were exclusively from the high-Ti group, and samples from the *Apollo*

12 and *15* and Luna 16 missions were from the low-Ti group. The age differences and the wide variety of lunar basalt chemistries implies that mare basalts cannot be generated from one common source region or common parental magma through different degrees of partial melting. Chemically, isotopically and mineralogically distinct source regions within the lunar interior are required (Figure 9.8). The broad chemical characteristics of mare lavas and highland rocks are depicted in Figure 9.9.

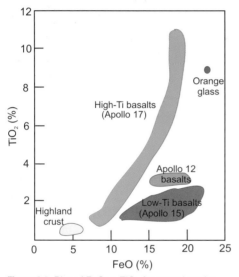

Figure 9.8 Plot of FeO vs TiO₂ for mare basalts, highland crust and orange glass. Apollo mission data. Courtesy NASA/JPL.

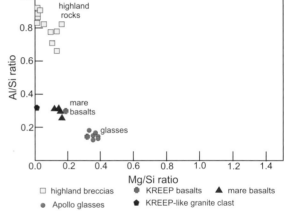

Figure 9.9 Plot of Al/Si vs Mg/Si for lunar basalts, glasses and highland breccias. Courtesy NASA/JPL.

Emplacement of mare lavas

Over large areas of the maria there is little evidence for flow fronts, probably because fissure-fed flood lavas were erupted at high rates over lengthy periods, covering their own fissures. However – and this is true particularly of late Imbrian to Eratosthenian aged flows in some maria (particularly in SW Mare

Imbrium) – spectacular lobate flow scarps can be identified (Figure 9.10). These flows cover an area of 200 000 km², often flow via braided channels, and in one case can be traced back to a source 370 km from the flow front. Many of the ridged flows have axial channels. The flows have an average thickness of 30 m.

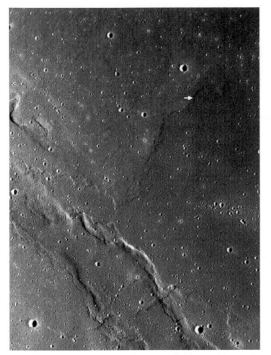

Figure 9.10 Lunar Reconnaissance Orbiter wide angle image of a part of Mare Imbrium. The low angle illumination brings out a flow front within the mare-fill lava sequence. Image field of view is 35 km. LROC WAC M177791761C per pixel. Courtesy NASA/GSFC/Arizona State University.

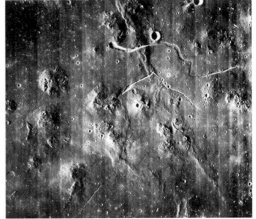

Figure 9.11 Lunar Orbiter 5 image of the region WNW of Marius crater. Note the prominent volcanic domes and cones and sinuous rilles. Image width 95 km. Frame M-215. Courtesy NASA/JPL/USGS.

Some of the most striking of mare landforms are the anastomosing sinuous rilles. These range in width between a few tens of metres to 30 km in width, the longest being 300 km in length. Most of these landforms are to be found near the borders of the maria, and as such are often associated with graben faults. This implies a connection with the extensional tectonics accompanying lava loading of the maria basins. They tend to cluster in groups, with which are often associated volcanic domes, depressions and small vents (Figure 9.11). Many rilles cross the lava plains as continuous open, flat-floored valleys but then break up into segments or rows of elongate pits. This, in many ways, is similar to terrestrial tube-channel systems, and the notion that these features represent major lava tubes or channels is now widely supported.

Other volcanic landforms

Volcanic domes tend to occur in clusters and may have summit depressions (Figure 9.11). Over 200 such structures have so far been identified. Some attain 20 km in diameter and may be 250 m high. Those nearer to the highland boundary tend to be more rugged than the very smooth-profile landforms typical of the mare surfaces. The latter seem to represent volcanic shields, indicative perhaps of lower rates of eruption towards the close of the voluminous plains-style activity. Their association with sinuous rilles adds weight to this idea.

Prominent on all of the maria are *en echelon* ridges, known as **wrinkle ridges** and broader lava arches**.** Often these have a concentric distribution with respect to the mare borders, but this is not the case everywhere. Individual ridges are several kilometres wide and reach heights of tens of metres. Various suggestions have been made to account for them: they may represent squeeze-ups produced during lava settling, low-angle thrust faults, or sit above laccolithic intrusions.

Lunar rocks and their radiometric ages

As is to be expected from the sequence of events on the Moon, the mare lavas are consistently younger than the highland rocks. Crystalline rocks from Mare Tranquillitatis have yielded K/Ar dates of 3.8Ga. Those from Oceanus Procellarum give K/Ar ages of 2.8Ga. *High-Ti basalts* collected from Mare Serenitatis and Mare Tranquillitatis yielded ages of 3.5–3.9Ga. *Low-Ti basalts* from Oceanus Procellarum and eastern Mare Imbrium gave ages of between 3.1 and 3.4Ga.

The highland samples yield greater ages. An anorthosite clast collected by *Apollo 15* yielded a Rb/Sr age of 4.42Ga, while a dunite clast collected by *Apollo 17*, using the same dating method, gave 4.47Ga. A norite breccia sampled by *Apollo 17* yielded a Rb/SR age of 4.33Ga. An *Apollo 16* plagioclase clast gave an Ar/Ar age of 4.33, and a Luna-20 anorthosite yielded 4.40Ga by the same method. One granitoid fragment sampled by *Apollo 17* gave a U/Pb age of 4.36Ga.

A brief lunar history

The first important event in the geologic evolution of the Moon was the crystallization of the near-global magma ocean, which may have been 500km thick. Differentiation of this ocean led to the flotation of less-dense feldspar crystals to the top, forming a layer perhaps 50km thick. The majority of the magma ocean crystallized quickly (<100my), although the final remaining KREEP-rich magmas, which are highly enriched in incompatible and heat-producing elements, could have remained partially molten for several hundred million years. The earliest phase of lava infilling of the nearside basins began about 4.2Gya, but the main phase occurred between 3.5 and 3.0 Gya, but continued until *c.* 2.0Gya.

Mascons and what they tell us

In the 1960s perturbations, induced on spacecraft by anomalously large mass concentrations within the moon's interior, had the potential to cause an orbiting craft to crash. These mass concentrations (mascons), are found beneath the surface of the Imbrium, Serenitatis, Crisium and Orientale impact basins, all of which possess prominent topographic lows and positive gravitational anomalies (Figure 9.12). Strangely, however,

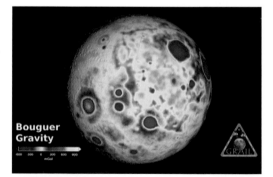

Bouguer
Gravity

Figure 9.12 Bougeur gravity anomaly map of the lunar nearside, showing mass concentrations (mascons) beneath the regular maria. Courtesy NASA/JPL-Caltech/MIT/GSFC.

the huge expanse of mare basaltic volcanism associated with Oceanus Procellarum does not possess one.

Theoretical considerations imply that a topographic low in isostatic equilibrium would exhibit a slight negative gravitational anomaly. Thus, the positive gravitational anomalies associated with these impact basins indicate that some form of positive density anomaly must exist within the crust or upper mantle that is currently supported by the lithosphere. Initially it was thought this could be accounted for by the basalt lava of the basin infill, which can be up to 6 km thick. However, while these lavas certainly contribute to the observed gravitational anomaly, uplift of the crust–mantle interface is also required to account for its magnitude.

Recent research suggests that during the early period of major bombardment, massive asteroids that collided with the moon left deep craters that reached into the mantle material that lies beneath the thin lunar crust. This caused surrounding lunar rocks from the moon's mantle to melt and collapse inward. This melting caused the material to become denser and more concentrated; upon cooling and crystallizing, this material became strong, so that it could support the load of highly dense material emanating from the mantle. The strong lunar crust, which also slid down into the impact cavity, eventually formed a curved but rigid barrier over the basin, holding the dense materials down.

Finally, Figure 9.13 provides a chemical comparison for a variety of volcanic rocks from all of the major bodies in the inner solar system, together with the Moon and carbonaceous chondrite meteorite. Figure 9.14 is a stunning image of Eugene Cernan beside a large rock at the *Apollo 17* site in Taurus-Littrow.

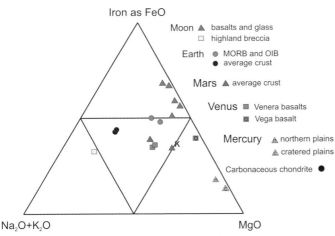

Figure 9.13 Mg/Fe/Na+K triangular diagram showing the chemical characteristics of a variety of volcanic rocks from the inner Solar System.

Figure 9.14 Photograph taken by Eugene Cernan, commander of *Apollo 17*, of lunar module pilot Harrison Schmitt standing in front of a large split boulder during the third EVA at the Taurus-Littrow site. The lunar rover is in the foreground at left. (Apollo 17, AS17-146-22294). Courtesy LPI/USRA.

10 The outer planets

The **frost line** marked an important dividing boundary within the solar nebula. Inside it temperatures were too high for hydrogen ices to form, so the only solid particles were made of metal and rock. Beyond the frost line, where hydrogen compounds could condense, the solid particles included ices as well as metal and rock (Figure 10.1). The outer planets, therefore, accreted their cores rapidly into large clumps of ice and rock. They then became so massive that they were able to capture a large amount of hydrogen and other gases from the nebula.

Jupiter and Saturn both formed when the Solar System was enveloped in gas. Once their nuclei were between 10 and 20 Earth radii

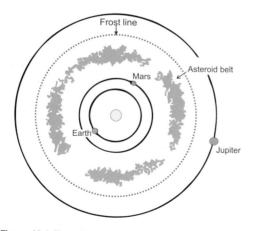

Figure 10.1 The 'frost line', located between the main asteroid belt and Jupiter, was the critical line beyond which hydrogen compounds could condense.

the surrounding gas could no longer remain dispersed in the nebula, therefore the growing planets attracted huge quantities of the gas. At the greater distances of Uranus and Neptune there was a battle between the accretion of the nuclei that triggered planetary growth and the dissipation of gas within the nebula. Because solid particles would have been more dispersed at these distances and orbital periods longer, it took longer for a sizeable body to aggregate. It appears that the nuclei lost out to the escape of nebular gas.

The cores of all four outer planets are made of a combination of rock, metal and hydrogen compounds, extending outwards into space. Jupiter and Saturn have similar interiors, with layers of metallic hydrogen, liquid hydrogen and gaseous hydrogen that are topped with a layer of visible clouds. Uranus and Neptune, on the other hand, have cores of rock and metal, but also water, methane and ammonia. The layer surrounding their cores is made of gaseous hydrogen, covered with a layer of visible clouds. The distinction of Jupiter and Saturn as 'gas giants' and Uranus and Neptune as 'ice giants' is very valid.

Although the radii of Jupiter and Saturn are very close, Jupiter is more than three times as massive. Jupiter has an average density of 1,326 kg/m^3, while Saturn's density is only 687 kg/m^3. This is because Jupiter continued to collect hydrogen and helium in its atmosphere, increasing its mass and compressing its interior until its average density was twice that of Saturn. Even though Neptune and Uranus are smaller in size, their densities are much larger than Saturn's (Neptune $1638\,kg/m^3$; Uranus $1270\,kg/m^3$). This leads us to believe that Uranus and Neptune must have a much larger fraction of higher-density material such as hydrogen compounds and rock.

Jupiter

Jupiter is much the largest and most massive planet in the Solar System; its mass is greater than those of all the other planets combined. The surface is dominated by the dark belts and bright zones, all of which are variable, although in general their latitudes change very little (Figure 10.2). There are also various striking features, notably

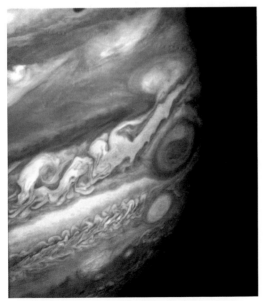

Figure 10.2 Voyager 2 image of Jupiter from a distance of 24 million km. Jupiter's moon Io can be seen at the right. Note the Great Red Spot and two prominent white ovals. P-21714. Courtesy NASA/JPL.

the Great Red Spot, but periodically brown ovals and white spots develop. Outbreaks in the South Equatorial Belt generally are the most spectacular phenomena seen; they involve sudden outbreaks of bright and dark clouds, with intense turbulence. Generally speaking, the North Temperate and South Temperate belts are well marked.

Jupiter's cloud layers are only about 50 km thick. Under them, when the pressure of the interior becomes high enough, the hydrogen of which Jupiter is made changes to liquid, which gradually changes further to liquid metallic hydrogen. The core is made out of heavier, rocky and metallic elements.

The inner layers of highly compressed hydrogen are in a state that has never been produced on the Earth. Normally, hydrogen does not conduct heat or electricity very well; thus, under normal conditions, hydrogen is not a metal. Under the extreme pressure found deep inside Jupiter, theory suggests that the electrons are released from the hydrogen molecules and are free to move about the interior. This causes hydrogen to behave as a metal: it becomes conducting for both heat and electricity. The intense magnetic field of Jupiter is thought to result from electrical currents in this region of metallic hydrogen that is spinning rapidly and thought to compose 75% of the planet's mass. The core is probably at a temperature as high as 24 000 K. It is small relative to the planet, about 20% of its radius, but it is still 15 times heavier than the Earth.

Jupiter radiates 1.6 times as much energy as falls on it from the Sun. Thus it must have an internal heat source. It is thought that much of this heat is residual heat left over from the original collapse of the primordial nebula, but some may come from slow contraction. This internal heat source is presumably responsible for driving the complex weather pattern in its atmosphere.

Undoubtedly the Great Red Spot is the most famous feature on Jupiter. Together with its characteristic 'hollow', it has certainly been in existence for many centuries. It is a phenomenon of Jovian meteorology, being a high-level anticyclonic vortex, with wind speeds of up to 360 km h⁻¹. To the south, the spot is encircled by an east wind, while to the north it is bounded by a strong west wind. This means that as the winds are deflected around it, they set up an anti-clockwise rotation, with a period of 12 days at the outer edge. The vortex is a high-pressure area elevated 8 km above the adjacent cloud deck by the upward convection of warmer gases from below.

The Galileo probe

After flybys of Earth, Venus and the asteroid belt, the Galileo spacecraft approached its final destination in 1994 and returned images of the comet Shoemaker-Levy 9 crashing into Jupiter. It made 35 orbits of the planet during its 8-year mission. A parachute probe, which separated from the orbiter in 1995, gathered data on the planet's turbulent atmosphere as it descended. Galileo continued to orbit Jupiter and make close flybys of its main moons until 2003.

Shoemaker-Levy collision

Between 16 July and 22 July 1994, fragments of Comet P/Shoemaker-Levy 9 collided with Jupiter, with dramatic effects (Figure 10.3). This was the first collision of two Solar System bodies ever to be observed. Shoemaker-Levy 9 consists of 20 fragments with diameters estimated at up to 2 km, which impacted the

Figure 10.3 A composite HST image showing the approach of Comet Shoemaker-Levy 9. The 21 fragments the fragments collided with Jupiter between 16 and 22 July 1994. Courtesy NASA, ESA, H. Weaver and E. Smith (STScI) and J. Trauger and R. Evans (NASA's Jet Propulsion Laboratory).

planet at 60 km/s. The impacts resulted in plumes many thousands of kilometres high, hot 'bubbles' of gas in the atmosphere, and large dark 'scars' on the atmosphere, which had lifetimes at least in the order of weeks.

Galileo detected a fireball which reached a peak temperature of about 24 000K, compared to the typical Jovian cloud-top temperature of about 130 K, before expanding and cooling rapidly to about 1500 K after 40 seconds. The plume from the fireball quickly reached a height of over 3 000 km. Shortly after the first impact, Earth-based observers saw a huge dark spot which was about 6 000 km across. This and subsequent dark spots were thought to have been caused by debris from the impacts. Over the next six days, 21 distinct impacts were observed, with the largest coming on 18 July when one created a giant dark spot over 12 000 km across, and was estimated to have released an energy equivalent to 6 000 000 megatonnes of TNT.

Galileo spectroscopic studies revealed absorption lines in the Jovian spectrum due to diatomic sulphur (S_2) and carbon disulphide (CS_2), the first detection of either in Jupiter, and only the second detection of S_2 in any astronomical object. Other molecules detected included ammonia (NH3) and hydrogen sulphide (H_2S). The amount of sulphur implied by the quantities of these compounds was much greater than the amount that would be expected in a small cometary nucleus, showing that material from within Jupiter was being revealed. As well as these molecules, emission from heavy atoms such as iron, magnesium and silicon was detected, with abundances consistent with what would be found in a cometary nucleus. Relatively low levels of water were later confirmed by Galileo's atmospheric probe.

Jupiter's ring system

Jupiter's ring was discovered by Voyager 1 and then confirmed by Voyager 2, which

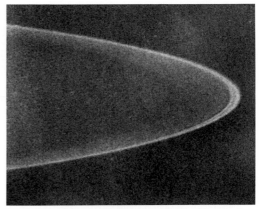

Figure 10.4 This processed image shows a narrow ring, about 1000 km wide, with a fainter sheet of material inside it. The faint glow extending in from the ring is likely to be caused by fine dust that diffuses in toward Jupiter. PIA9249. Courtesy NASA/Johns Hopkins University Applied Physics Laboratory/Southwest Research Institute.

obtained a set of definitive images (Figure 10.4). The ring is now known to be composed of three major components. The main ring is about 7 000 km wide and has an abrupt outer boundary 129,130 km from the centre of the planet. It encompasses the orbits of the two small moons, Adrastea and Metis. These may act as the source for the dust that makes up most of the ring. At its inner edge the main ring merges gradually into the 'halo.' The halo is a broad, faint torus of material about 20 000 km thick that extends halfway from the main ring down to the planet's cloud tops.

Saturn

Saturn is generally considered to be the Solar System's most beautiful object, largely on account of its magnificent ring system (Figure 10.5). It exhibits a yellowish hue due to ammonia crystals in the upper atmosphere. This atmosphere contains relatively

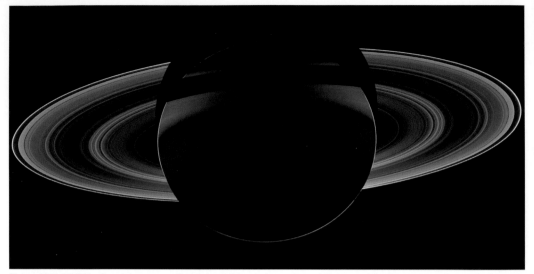

Figure 10.5 NASA's Cassini spacecraft obtained this glorious view of Saturn, while the spacecraft was in Saturn's shadow. PIA14934. Courtesy NASA/JPL-Caltech/SSI.

more hydrogen and less helium than Jupiter. Saturn is colder than Jupiter, so that ammonia crystals form at higher levels, covering the planet with 'haze' and giving it a somewhat bland appearance. Features over 1000 km in diameter are uncommon, and even the largest ovals are no more than half as big as Jupiter's Great Red Spot (Figure 10.5).

Saturn has a very hot interior, reaching 11,700°C at the core. Calculations made in 2004 indicated that the core must be between 9 and 22 times the mass of the Earth, which corresponds to a diameter of about 25 000 km. This is surrounded by a thicker liquid metallic hydrogen layer, followed by a liquid layer of helium-saturated molecular hydrogen that gradually transitions into gas with increasing altitude. The outermost layer stretches for 1 000 km and consists of a gaseous atmosphere.

The planet radiates 2.5 times more energy into space than it receives from the Sun.

Most of this extra energy is generated by slow gravitational compression, but this alone may not be sufficient to explain Saturn's heat production. An additional mechanism may be at play whereby Saturn generates some of its heat through the 'raining out' of droplets of helium deep in its interior. As the droplets descend through the lower-density hydrogen, the process releases heat by friction and leaves the outer layers of the planet depleted of helium. These descending droplets may have accumulated into a helium shell surrounding the core.

The Cassini-Huygens spacecraft images some tremendous storm vortices at Saturn's north pole. The angry eye of a hurricane-like storm appears dark red while the fast-moving hexagonal jet stream framing it is a yellowish green on Figure 10.6. Low-lying clouds circling inside the hexagonal feature appear as muted orange colour. A second,

Figure 10.6 Cassini probe image of a spectacular storm near Saturn's north pole. The view was acquired at a distance of 419 000 km. PIA.14946. Courtesy NASA/ JPL-Caltech/SSI.

smaller vortex pops out in teal at the lower right of the image. The rings of Saturn appear in vivid blue at the top right.

The ring system

The rings of Saturn have been given letter names in the order of their discovery. The main rings are – working outward from the planet – known as C, B, and A. The Cassini Division is the largest gap in the rings and separates Rings B and A. In addition, a number of fainter rings have been discovered more recently. The D Ring is exceedingly faint and closest to the planet. The F Ring is a narrow feature just outside the A Ring. Beyond that are two far fainter rings named G and E. The rings show a tremendous amount of structure on all scales; some of this structure is related to gravitational perturbations by Saturn's many moons, but much of it remains unexplained.

The ring particles are made almost entirely of water ice, with a trace component of rocky material. The particles vary from microns to a few metres in size. Some features of the rings suggest a relatively recent origin, but theoretical models indicate they are likely to have formed early in the Solar System's history.

The dense main rings extend from 7000 km to 80 000 km above Saturn's equator and have an estimated local thickness of as little as 10 metres and as much as 1 kilometre. Based on Voyager observations, the total mass of the rings was estimated to be about 3×10^{19} kg.

While the largest gaps in the rings, such as the Cassini Division and Encke Gap, can be seen from Earth, both Voyager 1 and 2 discovered that the rings have an intricate structure involving thousands of thin gaps and ringlets. This structure is thought to arise partly from the gravitational pull of Saturn's many moons, and also from resonances between the orbital period of particles in the gap and that of a more massive moon further out. For instance, the moon Mimas maintains the Cassini division (Figure 10.7).

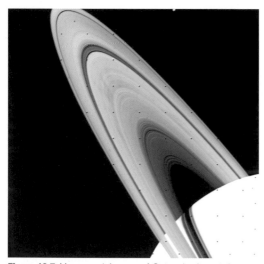

Figure 10.7 Voyager 1 image of Saturn's rings taken on 4 November 1980, eight days before closest approach. The structure of the rings is clearly visible in this raw clear-filter image. Courtesy NASA/JPL.

Data from the Cassini space probe indicate that the rings of Saturn possess their own atmosphere, independent of that of the parent planet. This is composed of molecular oxygen gas (O_2) produced when ultraviolet light from the Sun interacts with water ice in the rings. Chemical reactions between water molecule fragments and further ultraviolet stimulation create and eject, among other things, O_2. According to models of this tenuous atmosphere, H_2 is also present.

There are two main theories regarding the origin of Saturn's inner rings. One theory proposed that the rings were once a moon of Saturn whose orbit decayed until it came close enough to the Roche limit to be destroyed. The second theory is that the rings were never part of a moon, but are instead left over from the original nebular material from which Saturn formed. A more recent idea proposes that the rings could represent part of the remains of the icy mantle of a much larger, differentiated moon that was stripped of its outer layer as it spiralled into the planet during the formative period when Saturn was still surrounded by a gaseous nebula.

Uranus

Like the atmospheres of Saturn and Jupiter, Uranus's exhibits a striped pattern of winds that circle its longitudes; however, it is cloaked by an envelope of methane. It is the absorption of red wavelength light by this methane layer that gives Uranus its turquoise appearance. At greater elevation in the stratosphere hovers a hydrocarbon smog. The composition of Uranus's atmosphere is 82.5% molecular hydrogen, 15.2% helium, and 2.3% methane, together with trace amounts of **hydrogen deuteride** (HD) and possible aerosols of ammonia ice, water ice, ammonia hydrosulphide, and methane ice.

It is likely that Uranus has a core of rock somewhere between the sizes of Earth and Earth's moon. The planet's interior is primarily composed of methane ice. Unlike both Jupiter and Saturn, Uranus appears to have a uniform composition with no internal layering. Because of its great distance from the Sun, Uranus actually radiates more heat than it receives, due to the release of heat from the convection of liquid hydrogen in its core. Since Uranus 'lies on its side', its south pole is the closest to the sun during a part of its orbit, thus causing the polar regions to absorb more energy than the equatorial ones. [

The Voyager 2 spacecraft discovered Uranus's faint rings (Figure 10.8). Two of the moons that the probe discovered, Cordelia and Ophelia, play a part in maintaining them. They are 'shepherd' moons that keep the constituent particles of the ring in tight order.

Neptune

The first detailed information about Neptune came from Voyager 2 in 1989. Neptune may be a twin of Uranus but it is a non-identical twin. It is appreciably denser and therefore more massive. It also has an internal heat source, for it sends out 2.6 times more energy than it would do if it depended entirely upon insolation. It does not share in Uranus's unusual axial tilt; at the time of the Voyager 2 pass it was Neptune's south pole that was in sunlight.

Like Uranus, it is a planet dominated primarily by ices. It seems that Neptune has a core made up of iron, nickel and silicates, at a temperature of at least 5000°C; the core mass

Figure 10.8 A Hubble Space Telescope image of Uranus surrounded by its four major rings and by 10 of its 17 known satellites. PIA02963. Courtesy NASA/JPL/STScI.

is thought to be around 1.2 times that of the Earth. The core may or may not have a well-defined boundary, but it probably extends out to around one-fifth of the planet's radius. Surrounding it is the liquid mantle, rich in water, methane, ammonia and other ices; the total mass of the mantle is between 10 and 15 times that of the Earth, and the temperature is high, so that the mantle is sometimes called a 'water–ammonia ocean'.

At the time of the Voyager 2 encounter, the most conspicuous feature on Neptune was a huge oval, the Great Dark Spot, with a longer axis of about 10 000 km; it lay at latitude 22°S, had a rotation period of 18.296 h, and drifted westward at about 325 m s^{-1} relative to the adjacent clouds. Evidently it was a high-pressure area, rotating counter-clockwise and showing all the characteristics of an atmospheric vortex. Hanging above it were bright cirrus-type clouds, made up of methane ice (Figure 10.9).

Neptune also has a rather obscure ring system, which was imaged first by the Voyager spacecraft by the occultation method. The rings have a rather 'clumpy' structure, and the outermost ring extends out to a distance of 62,900 km from the primary. The albedo is very low indeed; they are about as reflective as soot! (Figure 10.10.)

Figure 10.10 These two 591-second exposures of the rings of Neptune were taken with the clear filter by the Voyager 2 wide-angle camera on 26 August 1989 from a distance of 280,000 km (175,000 mi). The two main rings are clearly visible and appear complete over the region imaged. PIA0 1997. Courtesy NASA/JPL.

Figure 10.9 Voyager 2 image of Neptune, showing the Great Dark Spot (GDS) which is about 13,000 km by 6600 km in size – as large along its longer dimension as the Earth. The bright, wispy 'cirrus-type' clouds seen hovering in the vicinity of the GDS are higher in altitude. PIA02245. Courtesy NASA/JPL.

Outer Solar System dynamics

After the formation of the Solar System, the orbits of all the giant planets continued to change slowly, influenced by their interaction with the large number of remaining planetesimals. About 4 billion years ago, Jupiter and Saturn fell into a 2:1 orbital resonance, Saturn orbiting the Sun once for every two Jupiter orbits. This resonance created a gravitational push against the outer planets, causing Neptune to travel past Uranus and interact with the dense belt of planetesimals.

The planets scattered the majority of the small icy bodies inwards, while they themselves moved outwards. This process continued until the planetesimals interacted with Jupiter, whose huge gravitational pull sent them into highly elliptical orbits or even ejected them from the Solar System. This caused Jupiter to move slightly inward. The outer two planets of the Solar System, Uranus and Neptune, are believed to have migrated outward in this way from their formation in orbits near Jupiter and Saturn to their current positions, over hundreds of millions of years.

11 Outer planet moons – Jupiter and Saturn

All of the gas and ice giants have natural satellites; twenty three of these have diameters that exceed 200 km. Jupiter has 67 known moons with secured orbits. Its eight regular moons are grouped into the planet-sized Galilean moons and the far smaller Amalthea group. Saturn has 62 moons with confirmed orbits, most of which are quite small; the seven larger moons include Titan, the second largest moon in the Solar System. The Saturnian rings are made up of icy objects ranging in size from one centimetre to hundreds of metres, each of which is on its own orbit about the planet. At least 150 small satellites or 'moonlets' embedded in the rings have been detected, and there may be many more.

How did the moons form?

With the exception of Triton, all of the outer planet moons orbit in the plane of their mother planet. Had they been captured, this arrangement is extremely unlikely, therefore it has to be assumed that they developed out of disks of particles surrounding their primaries. There is uncertainty about Neptune, as its large moon, Triton, has a retrograde path and is also inclined at 21° to Neptune's orbital plane. It is more likely that Triton was captured (Figure 11.1).

The problem with Uranus lies in the fact that its rotational axis is tilted at 98° to the plane of its orbit. The simplest way to account for this situation is to presume that, at an early stage in the development of the Solar System,

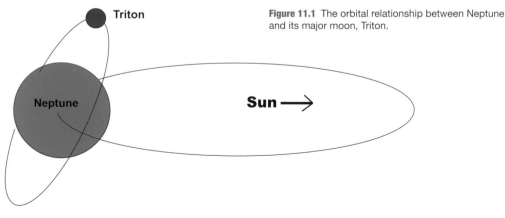

Figure 11.1 The orbital relationship between Neptune and its major moon, Triton.

Uranus collided with a large planetesimal, causing it to be tilted 'on end'. Had the satellites we see now existed then, it is very unlikely that they would have remained all in the same plane, therefore they may have accreted from the particles surrounding Uranus after the impacting body had been disrupted.

Composition of outer planet moons

Whereas the inner planet moons are made up largely of silicates, the great majority of the larger outer moons are a mixture of ice and rock. Their densities indicate that there is about 60% ice and 40% rock – roughly the cosmic abundance of these substances. Initially they were a homogeneous mixture of ice and rock. The majority of the ice present is water, but ammonia and methane also occur. These have the effect of lessening the density, while ammonia also lowers the melting temperature of ice.

The more massive moons almost certainly would have been warmed sufficiently during the accretional stage to have melted, which would have enabled the denser silicate particles to sink towards their cores, forming differentiated bodies with an outer layer made up largely of water, which eventually froze (Figure 11.2).

The decay of long-lived radioactive nuclides, at least for moons with diameters of >600 km, would have provided sufficient

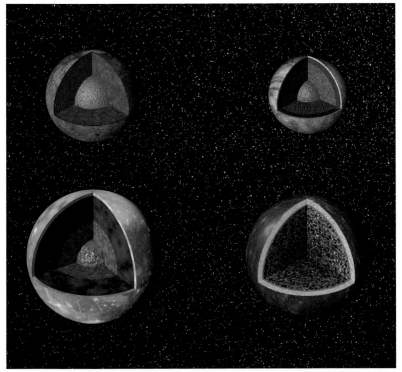

Figure 11.2 The interiors of five of the larger outer planet moons. Considerable differences in structure arise from variations in history and location. Courtesy NASA/JPL.

thermal energy to induce melting. Larger worlds, however, would have formed a silicate core and, depending on the rate of dissipation of the heat provided by radioactive decay, would have melted the rock–ice-mixture remaining, from the inside outwards. This could lead to a moon with a rocky core, encased in a layer of ice which, in turn, was mantled by an outer layer composed of both rock and ice.

A further source of planetary heat is produced by tidal effects. If a moon's orbit is not completely circular, there will be continuous distorting of the body and the generation of a tidal bulge. Any moon with an even slightly elliptical orbit will experience speeding up and slowing down; and the forces required to keep any bulge in its shape facing the planet around which it rotates warm the moon's interior.

Cryovolcanism

Once spacecraft explored the moons of the outer Solar System, the term 'volcanism' had to be extended to include the activities of magmas that were not composed solely of silicates. This would include sulphur and its compounds, ice, ammonia compounds and methane. **Cryovolcanism** may therefore include any of these materials, plus silicates rising from greater depth within these bodies. Ice-volcanic melt is the fluid or semi-fluid material associated with ice volcanism; this can have a wide range of rheology.

Jupiter's Galilean moons

Jupiter's four major moons, known as the Galilean Figure 11.1 The orbital relationship between Neptune and its major moon, Triton. moons, are all over 3100 km in diameter, and Ganymede is actually larger than the planet Mercury (Figure 11.3). The three innermost

Figure 11.3 Jupiter's four large 'Galilean' satellites as seen by the Long Range Reconnaissance Imager (LORRI) on the New Horizons spacecraft during its flyby of Jupiter in late February 2007. The four moons are, from left to right: Io, Europa, Ganymede and Callisto. PIA09352. Courtesy NASA/John Hopkins University.

moons – Io, Europa and Ganymede – share a 1:1, 2:1 and 4:1 orbital resonance. This is due to the gravitational tidal effects of Jupiter on the satellites and the satellites on each other. Since Io and Europa are not always at the same distance from Jupiter, the rate at which they orbit also changes, and thus they cannot keep the same face to Jupiter all the time. Jupiter's tidal force therefore squeezes each of them and warms them far beyond a level they could sustain normally. To a lesser extent the same is true of Ganymede.

Callisto, the outermost and least dense of the four moons, has a density intermediate between ice and rock, whereas Io, the innermost and densest moon, has a density intermediate between rock and metallic iron. The three inner moons have differentiated interiors; furthermore, each shows significant alteration at the surface (Figure 11.4).

The nearer a moon is to Jupiter, the hotter will be its interior. With the exception of Callisto, this will have melted the interior ice, allowing rock and iron to sink towards the core and water to cover the surface. In the case of Io, the heating was so extreme that all the rock has melted and water has long ago boiled off into space. As a result, Io is the most volcanically active object in the Solar System.

Io

The Voyager spacecraft took the first close-up images of Io. These showed a surface almost entirely covered with large volcanoes and no signs of impact craters. Some of the volcanoes were actually erupting as Voyager passed. The frequency of these sulphur-rich eruptions has filled in almost all of the impact craters and left Io with one of the youngest looking surfaces in the solar system (Figure 11.5).

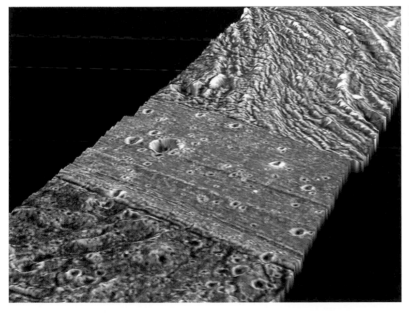

Figure 11.4 NASA's Galileo spacecraft took this image of Arbela Sulcus, a 24 km-wide region of furrows and ridges on Jupiter's moon Ganymede. PIA02576. Courtesy NASA/JPL/Brown University.

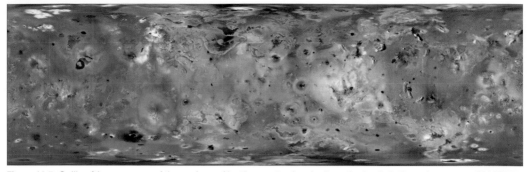

Figure 11.5 Galileo/Voyager map of the surface of Io, the most volcanically active body in the solar system. PIA09257. Courtesy NASA/JPL/USGS.

Close-up photos of eruptions in progress show powerfully hot lava glowing orange and red. Images obtained on the night side show not only the hot volcanic vents, but also a thin sulphur dioxide atmosphere produced by constant outgassing. Io's unusual red and orange colours come primarily from different **allotropes** of sulphur, which condenses on the surface after being outgassed by the volcanoes.

Although there is no direct evidence of tectonic activity, scientists feel confident it exists, since the processes that fuel volcanism also fuel tectonics. The volcanic eruptions are so frequent and cover the surface so thoroughly that any clear evidence of tectonic activity is likely to be buried.

Three kinds of terrain exist on Io: (i) mountain material, (ii) plains, and (iii) vents. Mountain material accounts for a mere 2% of the surface and probably represents outcrops of volcanically-generated silicate rocks that have been mantled in sulphur-rich sublimates. (Figure 11.6).

The layered plains are the predominant landforms, and range in hue from black, through orange and yellow to white. About

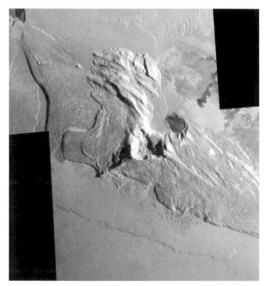

Figure 11.6 Spectacular Galileo image of Io's mountainous landscape around a peak named Tohil Mons. It rises 5.4 km above Io's surface. PIA03600. Tohil Mons rises 5.4 km above Io's surface, according to analysis of stereo imaging from earlier Galileo flybys of Io. Courtesy NASA/JPL/University of Arizona.

40% of these units have intermediate albedo and smooth surfaces and are considered to represent the ballistic fallout from volcanic eruptions, or volcanic flows interbedded with fumarolic products (Figure 11.7). A smaller area of plains, also with smooth surfaces, is incised by steep scarps that may exceed heights of 1 kilometre. This strongly suggests that there has been headward erosion of the units, possibly by sapping or wall slumping along pre-existing faults.

Over 400 active volcanic vents have now been mapped. Several produce plumes of sulphur and sulphur dioxide that climb as high as 500 km. The materials produced by this volcanism make up Io's thin, patchy atmosphere, while the volcanic ejecta produce a large plasma torus around Jupiter.

The eruption plumes are of two main types. (1) Short-lived plumes that rise to heights of 300 km and deposit dark deposits up to 1500 km in diameter; an example of this type is

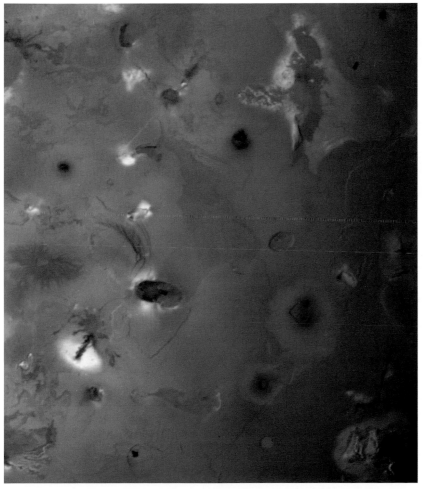

Figure 11.7 Typical plains units near to Io's southern pole. They are a mixture of volcanic centres and flat-lying, stratified deposits of volcanic origin. PIA01485. Courtesy NASA/JPL/ USGS.

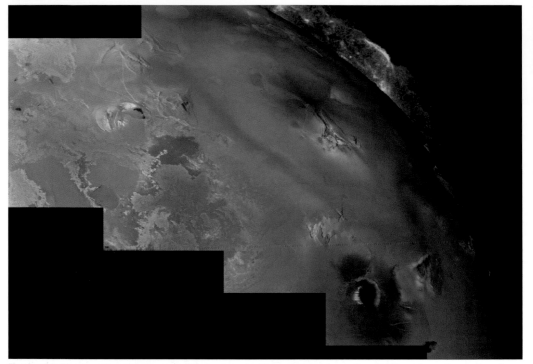

Figure 11.8 The eruption of Pele on Jupiter's moon Io. The volcanic plume rises 300 km above the surface in an umbrella-like shape. The plume fallout covers an area the size of Alaska. The vent is a dark spot just north of the triangular-shaped plateau (right centre). To the left, the surface is covered by colourful lava flows rich in sulphur. PIA00323. Courtesy NASA/JPL/University of Arizona.

the plume of Pele (Figure 11.8). (2) Long-lived plumes reach lesser heights (50–120 km) and spread paler deposits that extend for 300 km; Prometheus is the type example (Figure 11.9).

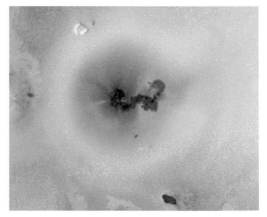

Figure 11.9 The plume at the volcano Prometheus. The long-lived plume has produced a ring-like deposit of bright white and yellow material that is likely to be rich in sulphur dioxide frost. Also note the denser jets in the plume that point like spokes to its source. Galileo scientists do not yet know whether this long-lived plume is erupting from a vent at the west end of the lava flow, or if the plume is being produced by the advancing lava as it flows over ground rich in sulphur dioxide. PIA 50596. Courtesy NASA/JPL/University of Arizona.

Temperatures at Prometheus-type vents are < 400K, while those associated with Pele-style plumes are higher, at 600–700K.

The plethora of lava flows that spread out across the Ionian surface are usually highly coloured. This poses the question of whether they represent silicate flows coated in sulphur compounds, or are molten sulphur flows. Sulphur melts behave in a very different way from silicate melts, for the latter show a marked increase in viscosity as they cool, whereas a pure sulphur flow would experience a dramatic decrease in viscosity (a thousand-fold decrease) as the temperature fell from 440 to 430K. They should, therefore, exhibit very different morphologies in the sense that long-run-out sulphur flows would suddenly fan out into broad pool-like landforms at their distal ends. While some flows do exhibit such morphology, very many do not. The jury is therefore out on this topic, but there is quite compelling evidence that silicate volcanism is an important process on this moon, as the prominent scarps and well-defined steep inner walls of the large calderas demand a stronger material than sulphur compounds to survive.

Europa

Europa's surface and crust are made almost entirely of water ice, and its bizarre, fractured appearance is proof enough that tidal heating has acted there. The icy surface is nearly devoid of impact craters and may be only a few million years old. Observations made by the Galileo spacecraft show that Europa has a metallic core and a rocky mantle. Surrounding the rocky interior appears to be an icy layer 100 km thick, the top few kilometres of which are frozen solid.

Data from the Galileo orbiter showed that Europa has an induced magnetic field through interaction with that of Jupiter, which suggests the presence of a subsurface conductive layer. The layer is likely to be a salty water ocean. The strong tidal forces likely provide enough heat to produce the subcrustal liquid water, in which case Europa may have an ocean containing more than twice as much liquid water as all of Earth's oceans combined.

Europa's most striking surface features are a series of dark streaks, or *lineae*, that criss-cross the surface; the larger bands are more than 20 km across (Figure 11.10). There is a strong likelihood that these landforms were generated by a series of eruptions of warm ice as the Europan crust opened to expose warmer layers beneath. The stresses are thought to have been caused in large part by the tidal stresses exerted by Jupiter.

Europa has a tenuous atmosphere composed mostly of molecular oxygen (O_2). The surface pressure is 10^{-12} times that of the Earth.

Ganymede

Ganymede is the largest moon in the Solar System, having a diameter of 5262 km and a density of 1.93 g/cm³. At least 50% must be water ice. The likelihood is that it has a 500 km-thick icy lithosphere below which is a 900 km-thick ice and silicate mantle. The core, around 2500 km in diameter, is likely to be silicate.

The surface of Ganymede shares many similarities with Europa. It is also made of water ice, but unlike Europa's surface, it shows signs of varying age. The darker regions are heavily cratered, suggestive of great age, while some of the lighter regions show no signs of craters,

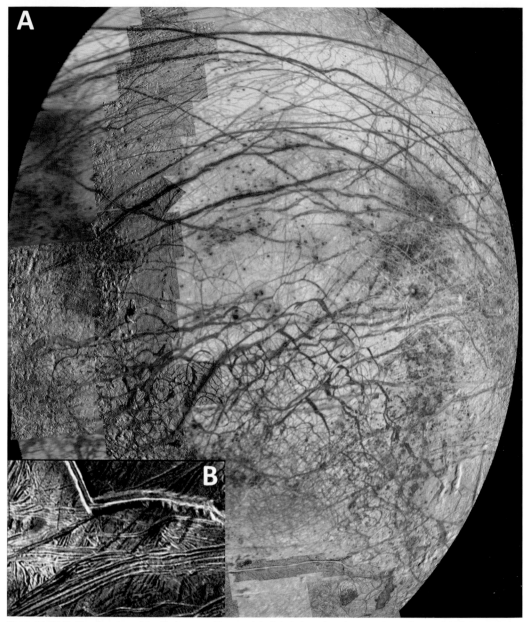

Figure 11.10 Galileo images of Jupiter's moon Europa. **(a)** global view showing prominent lineae. **(b)** High resolution view of lineae. PIA 17737/16827. Courtesy NASA/JPL/Caltech/University of Arizona.

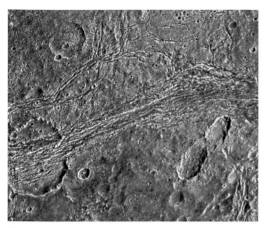

Figure 11.11 Galileo image of deformed dark terrain on Jupiter's moon, Ganymede. On the left is a crater that has been torn apart by tectonic forces. A lane of ridges and grooves (probably extensional fault blocks) cuts through the crater and distorts its originally circular shape. PIA01612. Courtesy NASA/JPL/Brown University.

and it is thought that eruptions of water covered the surface before freezing over. Other regions show a range of impact crater densities and spectacular deformational features (Figure 11.11).

Dark terrain (Figure 11.12) represents the oldest preserved surface on Ganymede. It is structurally modified by remnants of vast multi-ringed structures termed *furrow systems*, the oldest recognizable structures on the surface. The furrows have been interpreted as fault-induced troughs formed by large impacts into a relatively thin lithosphere early in Ganymede's history. Evidence suggests that the furrowed terrain represents a relatively thin and dark lag deposit of non-icy material, over a brighter, cleaner icy layer.

Light terrain forms swaths that can be subdivided into polygons that may be anything between 10 and 100 km across. They form an intricate patchwork across the surface and are characterized by sets of subparallel ridges and troughs.

Callisto

Callisto is the stereotypical outer Solar System satellite. It is one of the largest and most heavily cratered satellites in the Solar System. The surface is very icy and dates back 4 Ga. It has a brownish tinge, probably

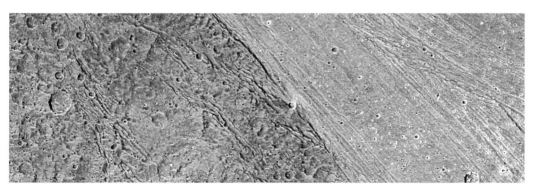

Figure 11.12 The ancient, dark terrain of Nicholson Regio (left) shows many large impact craters, and zones of fractures oriented generally parallel to the boundary between the dark and bright regions of Jupiter's moon Ganymede. In contrast, the bright terrain of Harpagia Sulcus (right) is less cratered and relatively smooth. PIA02577. Courtesy NASA/JPL/DLR.

Figure 11.13 Galileo complete global colour image of Callisto. Of Jupiter's four largest moons, Callisto orbits farthest from the giant planet and is the most heavily cratered. PIA 03456.Callisto's surface is uniformly cratered but is not uniform in colour or brightness. Scientists believe the brighter areas are mainly ice and the darker areas are highly eroded, ice-poor material. PIA 03456. Courtesy NASA/JPL/Brown University.

due to ejected materials from the myriad impact craters (Figure 11.13). Beneath the icy crust is possibly a salty ocean supported by a deeper rocky interior. Callisto lacks mountains, any evidence of volcanic or tectonic activity, or of any appreciable level of internal heat. Nonetheless, observations of Callisto's magnetic field may necessitate us having to add it to the list of possible worlds with subsurface salty oceans.

One of the most astounding features is the 4000-kilometre-diameter Valhalla impact basin, one of several such landforms. This consists of at least 25 concentric rings or ring arcs, and is one of the most unusual impact structures in the Solar System. The unusual number of rings may be due to impact into a thin, icy lithosphere.

The satellites of Saturn

Twenty-four of Saturn's moons are regular satellites; they have prograde orbits not greatly inclined to Saturn's equatorial plane. Sixteen moons keep the same face toward the planet as they orbit. Two of the moons orbit within gaps in the main ring system. The larger bodies show a wide range of geological features, including impact craters, grooves and cryovolcanic features. The large moon, Titan, has an atmosphere of nitrogen compounds that obscures its surface. It is so large (diameter 5150 km) that it affects the other moons of the Saturnian system. It contains 96% of the mass of the bodies orbiting Saturn.

The relatively large moon, Hyperion, is locked in a resonance with Titan. The remaining regular moons orbit near the outer edge of the A Ring, within the G Ring and between the major moons Mimas and Enceladus. The remaining 38, all small except one, are irregular satellites, whose orbits are much farther from Saturn, have high inclinations, and are mixed between prograde and retrograde. These moons are probably captured minor planets, or debris from the breakup of such bodies after they were captured. The largest of the irregular moons is Phoebe.

Cassini-Huygens

Cassini–Huygens is a NASA-ESA-ASI spacecraft that has studied the planet and its many natural satellites since arriving there in 2004, also observing Jupiter, the heliosphere, and testing the theory of relativity. Launched in 1997 after nearly two decades of development, it included a Saturn orbiter and an atmospheric probe/lander (Huygens probe) for Titan, which entered and landed on Titan in 2005.

On 14 January 2005, the Huygens probe entered Titan's atmosphere and descended to the surface. It successfully returned data to Earth, using the orbiter as a relay. This was the first landing ever accomplished in the outer Solar System.

Titan

Titan is the second biggest moon in the Solar System. Like Ganymede, it has a larger radius than Mercury, but a smaller mass because of its low density. At the large distance of Saturn, it is sufficient for it to retain an atmosphere, which in fact, is about 50% thicker than Earth's, although composed of 90% nitrogen and about 10% methane, with no oxygen. This is facilitated by the remarkably cold temperature (93 K or –180°C).

The composition in the stratosphere is 98.4% nitrogen, with the remaining 1.6% composed mostly of methane (1.4%) and hydrogen (0.1–0.2%). Titan's main ionosphere lies at an altitude of 1,200 km but there is an additional layer of charged particles at 63 km. Methane condenses out of Titan's atmosphere at high altitudes, so its abundance increases in descending below the tropopause at an altitude of 32 km, levelling off at a value of 4.9% between 8 km and the surface. The orange colour as seen from space may be produced by **tholins**.

Titan is likely differentiated into several layers with a 3,400 km rocky core surrounded by several layers composed of different crystal forms of ice (Figure 11.14). Its

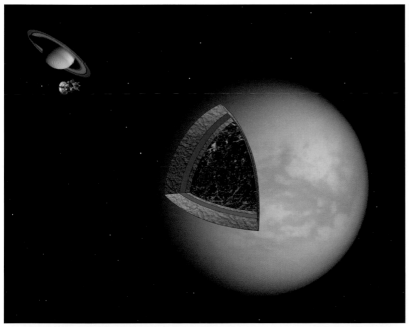

Figure 11.14 Artist's impression of the interior of Saturn's moon, Titan. PIA12843. Courtesy NASA/JPL.

interior may still be hot, and there appears to be a liquid layer composed of water and ammonia between the low pressure ice crust, and deeper ice layers made of high-pressure forms of ice. The presence of ammonia allows water to remain liquid even at temperatures as low as 176K (−97°C).

Despite being a cold world, the temperature and the atmospheric pressure are high enough for liquid methane to exist on the surface. In fact, when the Huygens probe landed on Titan, it returned pictures showing channel-like features that were probably carved by liquid methane.

During six flybys of Titan from 2006 to 2011, the Cassini spacecraft enabled scientists to infer Titan's changing shape. The density of Titan is consistent with a body that is about 60% rock and 40% water. However, the surface can rise and fall by up to 10 metres during each orbit; this degree of warping suggests that Titan's interior is relatively deformable, and that the most likely model of Titan is one in which an icy shell dozens of kilometres thick floats atop a global ocean that may lie no more than 100 km below its surface.

The first images revealed a diverse geology, with both rough and smooth areas, and light and dark regions. These include Xanadu, a large, reflective equatorial area about the size of Australia; morphologically complex, it is composed of hills and cut by deep valleys (Figure 11.15). In places the landscape is criss-crossed by dark lineaments sugges tive of tectonic activity. Several of the dark regions represent extensive plains covered

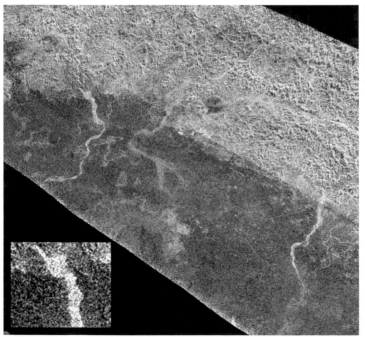

Figure 11.15 A Cassini-Huygens image showing the border of Xanadu as the bright–dark boundary running from the upper left to lower right. Southward from that boundary is an unusual set of channels. PIA10956. Courtesy NASA/JPL-Caltech/ASI.

in longitudinal sand dunes up to 330 m high, about 1 km wide, and tens to hundreds of kilometres long. Examination has also shown the surface to be relatively smooth; the few objects that seem to be impact craters appeared to have been filled in, perhaps by raining hydrocarbons. Radar altimetry suggests height variation is low, typically no more than 150 metres.

Near Titan's south pole, a peculiar dark feature named Ontario Lacus was identified as a lake. Radar images identified extremely smooth reflective areas in several places, which appear to be lakes filled with liquid methane. These are the first stable bodies of surface liquid found outside of Earth (Figure 11.16). Although most of the lakes are concentrated near the poles, a number of long-standing hydrocarbon lakes exist in the equatorial regions.

Some appear to have channels associated with liquid and lie in topographical depressions. Imagery also shows channels that have

created surprisingly little erosion, suggestive of the fact that on Titan erosion either is extremely slow, or some other phenomenon may have erased evidence of older riverbeds and landforms. During a flyby on 26 September 2012, Cassini's radar detected in Titan's northern polar region what appears to be a river system with a length of more than 400 km (Figure 11.17).

The few impact craters discovered include a 440 km-wide two-ring impact basin named Menrva, a smaller, 60 km wide, flat-floored crater named Sinlap, and a 30 km crater with a central peak and dark floor named Ksa.

In March 2009, structures resembling lava flows were announced in a region called Hotei Arcus. The putative flows were found to rise 200 m above Titan's surface. In December 2010, the Cassini mission team announced the identification of a major cryovolcano named Sotra Patera. It is one in a chain of at least three mountains, each between 1000 and

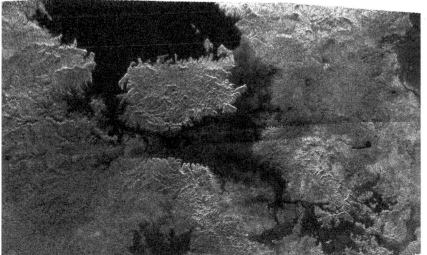

Figure 11.16 This Cassini-Huygens image shows an island or peninsula surrounded by lakes, close to the northern pole. Cassini-Huygens image of an island set amongst lakes near the north pole of Titan. PIA09180. Courtesy NASA/JPL/ESA/University of Arizona.

Figure 11.17 Ligeia Mare is the second largest known body of liquid on Saturn's moon, Titan. It is filled with liquid hydrocarbons, such as ethane and methane. River systems with typical tributaries can be seen quite clearly entering the lake. PIA17031. Courtesy NASA/JPL/ESA/University of Arizona.

1500 m in height, several of which are topped by large craters. The lower flanks appear to abut against frozen lava flows.

The moon's tough icy shell is apparently far stronger than previously thought. By comparing gravity results with the structure of Titan's surface, scientists were surprised to find that the regions of high elevation had the weakest gravitational pull, suggesting that such mountains have 'roots' somewhat like Earth's continents. Furthermore, these roots must be large enough to displace a lot of water under them, meaning they exert a weaker gravitational pull. Ice is buoyant in water, so in order to hold these big 'icebergs' down, the outer shell of Titan must be extremely rigid.

Mimas

The smallest and closest-orbiting of Saturn's major moons, Mimas, cleared the gap known as the Cassini division between two of the planet's rings. It is composed primarily of water-ice, but despite its proximity to the planet and the strong tidal stresses that must be felt, the surface of the moon appears to be unchanged. Its most striking feature is a 130 km-diameter crater, Herschel, away from which radiate a series of troughs (Figure 11.18).

Enceladus

The surface of this moon is morphologically diverse and includes ancient heavily cratered terrain as well as younger smooth areas. The cratered areas show the widest variation in crater density and form of any satellite in the Solar System. Its surface also has one of the highest albedos, and this may be due to a coating of fresh frost, possibly generated by cryogenic eruptions. Many plains units are fractured and intersected by grooves. The complexity of the units indicates that there have been several phases of resurfacing.

The narrower, straighter grooves, which may be up to 100 km long and typically 2–4 km wide, appear to represent fractures opened due to the freezing of water in the interior of the

Figure 11.18 Cassini Orbiter image of Mimas, showing Herschel Crater which is 130 km across. PIA12570. Courtesy NASA/JPL-Caltech/Space Science Institute.

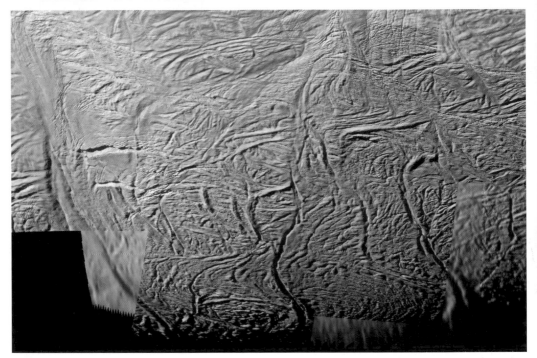

Figure 11.19 Cassini Orbiter mosaic Enceladus, showing tectonic deformation in the fractured south polar region, where jets of water ice spray outward to form Saturn's E ring. PIA11686. Courtesy NASA/JPL/Space Science Institute.

moon. The more curved grooves, which occur in swarms and may attain lengths of 200 km, are found on the younger ridged plains. Their origin is more equivocal; it has been suggested they are compressional features formed by the convergence of adjacent crustal blocks, but it has also been conjectured that they are graben faults (Figure 11.19).

The area around the moon's south pole was found to be unusually warm and cut by a system of fractures about 130 km long, some of which emit jets of water vapour and dust. These jets form a large plume above its south pole, which replenishes Saturn's E ring and serves as the main source of ions in the planet's magnetosphere. The gas and dust are released at a rate of more than 100 kg/s.

It appears that Enceladus may have liquid water underneath the south-polar surface. The source of the energy for this cryovolcanism is thought to be a 2:1 resonance with Dione, causing it to experience a tidal drag twice in every orbit of Dione. The fact that there have been several phases of volcanic resurfacing may well be a response to changing orbital resonances over time, activity switching on and off as these develop, fade and reform.

During the first two close flybys of the moon in 2005, Cassini discovered a 'deflection' in the local magnetic field that is characteristic for the existence of a thin but significant atmosphere. Other measurements obtained at that time point to ionized water vapour as being its main constituent. Cassini also observed water ice geysers erupting from the south pole, which gives more credibility to the idea that Enceladus is supplying the particles of Saturn's E ring. Mission scientists hypothesize that there may be pockets of liquid water near the surface of the moon that fuel the eruptions.

Rhea

Rhea is the second largest of Saturns' moons (diameter 15,284 km) and is heavily cratered. It is composed of ice, with between 35 and 40% of rock mixed in, causing it to resemble a dirty snowball; it appears to lack a core. It has a high albedo, apparently having a cleaner surface than Jupiter's heavily cratered moon Callisto. It has a very tenuous oxygen atmosphere, about 5 trillion times less dense than that of Earth, but the only known oxygen atmosphere in the outer Solar System. It is possible that radiation from Saturn's magnetosphere could release oxygen and carbon dioxide from the moon's icy surface.

The heavily cratered surface has two large impact basins on its anti-Saturnian hemisphere, measuring 400 and 500 km in diameter (Figure 11.20). There is also a 48 km-diameter impact crater called Inktomi which has an extended system of bright rays, making it very prominent, and also possibly one of the youngest craters on the inner moons of Saturn. The only other features of note are of a few large fractures on the trailing hemisphere.

Figure 11.20 Cassini Orbiter image of the heavily cratered surface of Rhea. PIA14605. Courtesy NASA/JPL/Space Science Institute.

Tethys

Tethys is the fourth largest of Saturn's inner moons (diameter 1058 km). A heavily cratered hilly terrain occupies the majority of its surface, while a less extensive, smoother plains region lies on the hemisphere opposite to that of the large impact crater Odysseus. The plains contain fewer craters and are apparently younger, while a sharp boundary separates them from the cratered terrain. A major chasm system called Ithaca Chasma extends for 270° around the moon's circumference. The chasm is concentric to the 400 km-diameter impact crater Odysseus, suggesting the two features may be related (Figure 11.21). There is also a system of extensional troughs radiating away from Odysseus. Tethys appears to have no current geological activity.

The moon's density ($0.985\,g/cm^3$) is less than that of water, indicating that it is made

Figure 11.21 Cassini Orbiter image of the crater Telemachus which sits within the northern reaches of Ithaca Chasma on Saturn's moon Tethys. Ithaca Chasma is an enormous rift that stretches more than 1000 km from north to south across the moon. PIA 10506. Courtesy NASA/JPL/Space Science Institute.

mainly of water ice with only a small fraction of rock. It travels close to Saturn and feels the gravitational pull of the planet. The heat from Saturn may allow the moon's icy surface to melt slightly, filling in craters and other signs of impact.

Hyperion

The smaller moon Hyperion is locked in a 4:3 resonance with Titan, whose nearest neighbour it is. Thus, while Titan makes four revolutions around Saturn, Hyperion makes exactly three. The moon has an extremely irregular shape, and a very odd, brownish icy surface. The average density of about 0.55 g/cm^3 indicates that the porosity exceeds 40% even assuming it has a pure ice composition.

The surface is peppered with numerous impact craters, those with diameters 2–10 km being particularly abundant. It is the only moon known to have a chaotic rotation. While on short timescales the satellite approximately rotates around its long axis at a rate of 72–75° per day, on longer timescales its axis of rotation wanders chaotically across the sky.

Iapetus

Iapetus is the third largest of Saturn's satellites and the most distant of the large moons. It has the greatest orbital inclination, at 14.72°. The moon has long been known for its unusual two-toned surface, the leading hemisphere being pitch-black and the trailing hemisphere almost as bright as fresh snow. The

Cassini probe showed that the dark material is confined to a large, near-equatorial area called Cassini Regio on the leading hemisphere; this extends from approximately 40°N to 40° S. The polar regions are as bright as its trailing hemisphere (Figure 11.22). Cassini also discovered a 20 km high equatorial ridge, which spans nearly the moon's entire equator. Otherwise both dark and bright surfaces of Iapetus are old and heavily cratered. The images revealed at least four large impact basins with diameters over 380 km.

A clue to the origin of the dark material comes from NASA's Spitzer Space Telescope, which in 2009 discovered a vast, nearly invisible disk around Saturn, just inside the orbit of the moon Phoebe. It is believed that the disk originates from dust and ice particles raised up by impacts on Phoebe. Because the disk particles, like Phoebe itself, orbit in the opposite direction to Iapetus, the moon collides with them as they drift in the direction of Saturn, coating its leading hemisphere.

Irregular moons

Irregular moons are small bodies with large-radii, inclined, and frequently retrograde orbits, believed to have been acquired by the parent planet through capture. They often occur as collisional families or groups. Phoebe is the outermost of Saturn's known moons, being almost four times more distant from Saturn than its nearest neighbour, Iapetus. Phoebe has a very low albedo (0.06), being as dark as lampblack. Its orbit is retrograde, inclined almost 175°, and is highly eccentric. It is probably a captured asteroid, like many of such moons.

Figure 11.22 Cassini Orbiter image of Iapetus's trailing hemisphere, broad tracts of which are almost as bright as snow. The dark terrain covers about 40 percent of the surface and is named Cassini Regio. Impact craters are abundant. PIA11690. Courtesy NASA/JPL/Space Science Institute.

12 Outer planet moons – Uranus and Neptune

Uranus has 27 named moons; of these, 13 inner moons orbit within Uranus's ring system, and another nine outside it; these are irregular. Neptune has 14 named moons; the largest, Triton, accounting for more than 99.5% of all the mass orbiting the planet but, oddly, having a retrograde orbit, suggesting it was captured. Neptune also has six known inner regular satellites, and six outer irregular satellites.

Of the dwarf planets, Pluto has five moons, its largest moon, Charon, being more than half as large as Pluto itself, and large enough to orbit a point outside Pluto's surface; in effect, they form a binary system. Pluto's four other moons are far smaller and orbit the Pluto–Charon system. The dwarf planet Haumea has two moons, Namaka and Hi'iaka, while Eris has one known moon, Dysnomia.

The satellites of Uranus

To date 27 moons are known to orbit this giant planet. The five large moons are composed of about 50% water ice, 20% carbon- and nitrogen-based materials, and 30% rock. Their surfaces, almost uniformly dark gray in colour, display varying degrees of geological activity. Voyager 2 obtained high-resolution images of each of Miranda, Ariel, Umbriel, Titania and Oberon. The two largest, Titania and Oberon, are about 1,600 km in diameter; the smallest,

Miranda, is only 500 km. The largest of the more recently detected moons, Puck, is 162 km in diameter: larger than most asteroids.

The satellites are divided into three groups: 13 inner moons, 5 major moons, and 9 irregular moons. The inner moons are small dark bodies that share common properties and origins with the planet's rings. Four of the five major moons exhibit signs of internally driven processes such as canyon formation and volcanism. Uranus's irregular moons have elliptical, strongly inclined and mostly retrograde orbits at great distances from the planet. The large moons are believed to have formed in the accretion disc that existed around Uranus in its early existence, or resulted from the large early impact suffered by Uranus.

Miranda

Miranda, the innermost of the five large moons, is one of the strangest bodies yet observed and appears to be composed largely of ice. Voyager images, which captured some areas of the moon at resolutions of a kilometre or less, revealed huge fault canyons as deep as 20 km, terraced layers and an amazing mix of old and young surfaces. The younger regions may have been produced by incomplete differentiation of the moon or, alternatively, Miranda may be a re-aggregation of material from a time when the moon

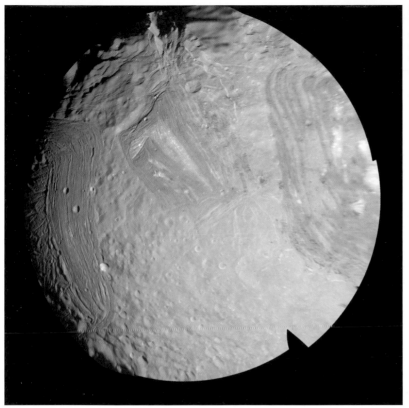

Figure 12.1 Voyager 2 image of the surface of Miranda, showing grooved terrain, cratered terrain and coronae. Courtesy NASA/JPL/Space Science Institute.

was smashed into pieces by a violent impact (Figure 12.1).

Roughly half of the moon's surface is of relatively uniform brightness and is densely cratered. Many of the craters have a less than sharp appearance, suggestive of mantling material that softens the topography. Others are far brighter and sharper in outline, and may reveal bright material in their inner walls; this can be hundreds of metres thick.

The remainder of the moon's surface is a complex mix of structures called *coronae*, composed of bands of sub-parallel ridges, amidst rolling terrain, and a relatively modest sprinkling of small bright craters, the largest of which is 300 km across. Elsinore corona, perhaps the least complex of such structures, has an inner region of fault-bounded blocks that is encircled by a zone of parallel ridges, each up to 1 km in height and several tens of kilometres long. The outermost part of the ridge belt appears to be the youngest and appears to sit on top of the cratered surface (Figure 12.2).

Inverness corona has a trapezohedral form and a more complex central region. As with Elsinore, it has a margin comprised in a ridge belt, but the central zone has odd bright

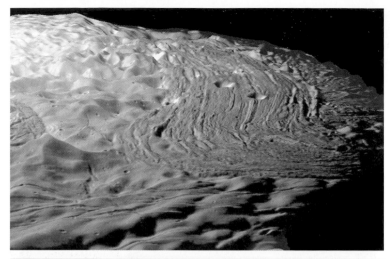

Figure 12.2 Voyager 2 image of Uranus's moon Miranda showing the 323 km diameter Elsinore Corona is the upper portion. This corona indicates vigorous tectonic activity at some point in Miranda's history. Courtesy NASA/JPL/Space Science Institute.

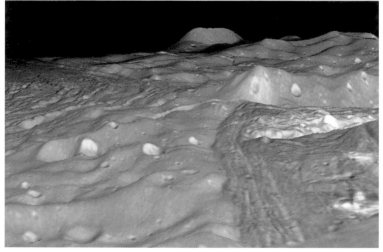

Figure 12.3 Voyager 2 image of Inverness corona, showing complex grooves and faulting. Note the many impact craters on the smoother hummocky terrain. Courtesy NASA/JPL/Space Science Institute.

albedo markings that are mimicked by a kind of ridge-and-furrow grain. Of great interest are the numerous steep fault scarps that cross the structure (Figure 12.3) and run into the adjacent cratered terrain. One of these soars to 10 km above the base of the scarp.

Arden corona is broadly similar to Elsinore. In the outer regions the normal faults, whose fault planes may attain 2 km in height, occasionally become tilted at very low angles (<20°) towards the outer margin and face outward, which presumably is a manifestation of significant extensional tectonics.

Explaining the geological relationships on this bizarre world is a major headache. One published theory suggests that the moon, which had partially differentiated, was disrupted and then re-accreted, probably as a random mixture of icy and rocky blocks. Subsequent diapiric activity, superimposed on a pre-existing regional stress system, then generated the coronae. The sets of intersecting fractures – many of which are graben – point to extensional forces, and these are most readily explained by assuming that Miranda expanded by a few percent, either by the freezing of a once-molten liquid zone, or by thermal expansion. The ridges probably represent extensional tilt blocks, the canyons graben formed by extensional faulting on top of diapirs or upwellings of warm ice. Other features seen on Voyager images may be due to cryovolcanic eruptions of icy magma.

Ariel

Ariel is the second closest large moon to the planet, orbiting at 190 000 km. Its density is 1.66 g/cm³, indicating that it consists of roughly equal parts water ice and a dense non-ice component. The presence of water ice is supported by infrared spectroscopic observations, which have revealed crystalline water ice on the surface. Water ice absorption bands are stronger on the leading hemisphere than on the trailing hemisphere, which may be related to bombardment by charged particles from Uranus's magnetosphere.

Ariel has the brightest and possibly the geologically youngest surface in the Uranian system, being largely devoid of craters greater than 50 km in diameter. It also appears to have undergone a period of even more intense activity, leading to many fault valleys and what appear to be extensive flows of icy material. Where many of the larger valleys intersect, their surfaces are smooth, which

could indicate that the valley floors have been covered with younger ice flows.

The main surface features are impact craters, canyons, fault scarps, ridges and troughs. The cratered terrain, which is the most extensive of the geological units, is to be found around the moon's southern pole. The largest crater observed is Yangoor, with a diameter of 78 km. It shows signs of subsequent deformation. Like all large craters on Ariel, it has a flat floor and central peak.

The cratered regions are intersected by a network of scarps, graben valleys (chasmata) and narrow ridges whose greatest concentration is in mid-southern latitudes. The chasmata generally are 15–20 km wide and usually have slightly convex floors. The longest, Kachina Chasma, is over 620 km long (Figure 12.4).

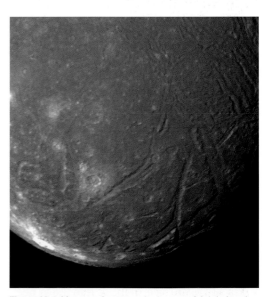

Figure 12.4 Voyager 2 composite image of Ariel showing intersecting valleys near the terminator. PIA00041. Courtesy NASA/JPL.

The ridged terrain comprises bands of ridges and troughs hundreds of kilometres wide, and cuts the cratered terrain into rough polygons. Within each ridge belt, which can be 70 km wide, are individual ridges and troughs up to 200 km long and between 10 and 35 km apart. The belts often form continuations of chasmata, suggesting that they may be a modified form of the graben or the result of a different reaction of the crust to the same extensional stresses.

Figure 12.5 Voyager 2 image of Titania showing impact craters and extensive faulted valleys. PIA00039. Courtesy NASA/JPL.

The youngest terrain is represented by the plains units. These are relatively low-lying smooth areas with a variable density of impact cratering that indicates a wide age range. Plains are found on the floors of canyons and within the cratered terrain, where they are separated by sharp boundaries. There are also some lobate landforms, indicative, perhaps, of formation by cryovolcanic processes.

Umbriel

Umbriel is ancient and dark, apparently having undergone little geological activity. Large craters pockmark its surface, but none have bright ejecta blankets or interior walls. The darkness of Umbriel's surface may be due to a coating of dust and small debris somehow created near and confined to the vicinity of that moon's orbit.

Titania

This moon is quite heavily cratered and has several basins <200 km in diameter. The younger craters have bright ejecta blankets, while the larger basins appear to have subsided somewhat into the moon's crust. Areas of lower crater density may have been resurfaced, probably by cryovolcanic activity, but possibly due to viscous relaxation. The surface is also marked by huge fault systems that attain heights of 5 km and may be as long as 1500 km (Figure 12.5). Opposing fault faces may form graben 20–50 km wide, and these often occur as multiple sets with intervening horsts. They are geologically young as they appear to transect all craters.

Oberon

The outermost of the pre-Voyager moons, Oberon, also has an old, heavily cratered surface with little evidence of internal activity other than some unknown dark material apparently covering the floors of many craters. However, it is different from Umbriel in that some craters have bright ejecta blankets. One image shows a 20 km high peak in profile on the moon's limb, indicating that the lithosphere must be thick and strong.

The satellites of Neptune

Neptune has 14 known satellites, the largest, Triton, being easily visible in a modest telescope. The seven moons which orbit inside Triton have prograde orbits and low eccentricities. Triton itself has retrograde motion and is almost certainly a captured body. The six outer irregular satellites, including

Nereid, whose orbits are much further from Neptune, have high inclinations, and are mixed between prograde and retrograde. The two outermost moons, Psamathe and Neso, have the largest orbits of any natural satellites discovered in the Solar System. All of the satellites have low albedo, presumably being made from ice with a coating of dark material, probably some form of carbon compound.

Triton

Triton follows a retrograde and quasi-circular orbit, and is thought to be a gravitationally captured satellite. It is the second known moon in the Solar System to have a substantial atmosphere, which is primarily nitrogen with small amounts of methane and carbon monoxide. The pressure on Triton's surface is about 14 μbar.

Triton is extremely cold, with a surface temperature of about 38 K (−235.2°C). Its highly reflective surface is covered by nitrogen, methane, carbon dioxide and water ices. Surface features include the large southern polar cap, older cratered plains cut by graben and scarps, and youthful features probably formed by cryovolcanism (Figure 12.6). Triton has a relatively high density of about 2 g/cm³, greater than that of both Ganymede and Callisto. This indicates that silicate rocks constitute about two-thirds of its mass, and ices (mainly water ice) the remaining one-third. There may be a layer of liquid water deep inside, forming a subterranean ocean. Voyager 2 observations revealed a number of active geysers within the polar cap when heated by the Sun; these eject plumes up to 8 km high. Impact craters occur, and these have a density similar to that of the lunar maria.

Figure 12.6 This Voyager 2 view of Triton shows a prominent chain of volcanic features surrounded by smooth volcanic plains, possibly formed by lavas or ash deposits of water or other ices, such as methane or ammonia. The smaller pits and domes are typically 10 km across and have relief of no more than a few hundred metres. PIA12185. Courtesy NASA/JPL/Universities Space Research Association/Lunar & Planetary Institute.

Voyager imagery revealed at least four terrain types. The most widespread units are (i) high smooth plains; (ii) hummocky plains; and (iii) cantaloupe plains. A fourth type is (iv) smooth floor material. The youngest unit appears to be the smooth plains, as it overlies the other two. In places it exhibits distinct lobate scarps up to a few kilometres high, and appears to have been emplaced as a high-volume viscous flow, perhaps a mixture of ammonia and water. The presence of circular depressions and lines of rimmed and rimless pits is pretty conclusive evidence for cryovolcanic activity.

Also apparently volcanic in origin is the smooth floor material. It outcrops in large depressions which may reach 200 km in diameter and appears to emerge from rough, pitted regions towards the centre of such depressions. The hummocky terrain is also likely of volcanic origin. It contains a large number of dome-like landforms and ridges.

Pluto
Hubble Space Telescope · Faint Object Camera

PRC96-09a · ST Scl OPO · March 7, 1996 · A. Stern (SwRI), M. Buie (Lowell Obs.), NASA, ESA

Figure 12.7 Voyager 2 view of Triton's cantaloupe terrain, most likely formed when the icy crust of Triton underwent wholesale overturn, forming large numbers of rising blobs of ice. The numerous irregular mounds are a few hundred metres high and a few metres across and formed when the top of the crust buckled during overturn. PIA12186. Courtesy NASA/JPL/Universities Space Research Association/Lunar & Planetary Institute.

Unique to Triton is the cantaloupe terrain. It comprises a close-knit mesh of roughly circular knobs between 5 and 25 km in diameter, crossed by a network of ridges that seem to represent extrusions from extensional fractures. It is on this terrain type that the majority of impact craters are to be found. Therefore, it is assumed to represent the oldest surface on the moon (Figure 12.7).

Voyager images revealed that a large area of Triton's southern hemisphere was covered in a bright polar cap which is presumed to be composed of nitrogen ice. This has a very ragged edge, suggestive of cap retreat. Voyager also recorded two eruptions on the cap, which sent material 8 km up into space, bending over at the tropopause and extending horizontally for over 100 km. It is possible that some of the dark streaks observed on the polar ice represent fallout from such eruptions. The suggestion that the plumes are akin to geysers has been made, but it also possible that enhanced midday solar energy was the cause.

Triton has an atmosphere that extends upwards to 800 km. This is composed mainly of nitrogen, but the surface pressure is a mere 14 microbars, which is 1/70 000th of the surface pressure on Earth. There are smaller amounts of methane and carbon monoxide, whose abundances are a few hundredths of a percent that of the nitrogen. In addition, the upper atmosphere contains significant amounts of both molecular and atomic hydrogen.

Proteus

Proteus, with a maximum diameter of 400 km, resembles an irregular polyhedron, with several flat or slightly concave facets 150 to 250 km in diameter. The surface is heavily cratered and shows a number of linear features. Its largest crater, Pharos, is more than 150 km in diameter.

Nereid

Nereid is the third largest moon of Neptune. It has a prograde but very eccentric orbit and is believed to be a former regular satellite that was shifted to its current orbit through gravitational interactions during Triton's capture. Spectroscopes have detected water ice on its surface; its spectrum being intermediate between Uranus's moons Titania and Umbriel.

Pluto and Charon

The dwarf planet Pluto and its largest satellite, Charon, are Kuiper Belt objects. Four much smaller bodies also form a part of this system. They were first imaged by the Hubble Space Telescope in 1994. The images showed that whereas Pluto has a reddish hue, Charon is a more neutral grey colour, indicative of different surface composition.

Charon is almost half as big as Pluto and rotates around it in 6.4 Earth days. It is tidally locked to its primary. The density of Pluto is about 2 g/cm^3 – corresponding to about 70% rock and only 30% ice – while Charon has a mean density of 1–2 gm/cm^3 and is presumed to be made largely of ice. Unlike Pluto, which is covered with nitrogen and methane ices, Charon's surface appears to be dominated by water ice, and also appears to have no atmosphere. In 2007, observations by the Gemini Observatory of patches of ammonia hydrates and water crystals on the surface of Charon suggested the presence of active icy geysers.

13 Epilogue

What of the future? Currently the highly successful Curiosity rover is conducting a variety of experiments and sending fascinating data back to Earth. A number of other recent missions are concentrating on comets, asteroids, Mars and the Moon. Others are planned for the future.

In late September of 2013 the Curiosity rover's science team reported that they had successfully dated a clay rock sample from within Gale crater. One set of results pinned down the actual age of the sample, while another pinpointed how long the rock had been exposed at the surface of the planet. Surprisingly the sample date was >3860 million years and the exposure age between 48 and 108 million years. The former must presumably also represent the age of the Gale impact.

One of the big questions about clays on Mars is: where did they form? Did they form like most clays on Earth, where abundant water on the surface attacked silicate minerals, altering them, eroding them and transporting them elsewhere? Or did they form in situ, by groundwater alteration of the silicate materials? As yet this is unanswered.

Interestingly, in terms of bulk chemistry (the amount of each chemical element present in the rock), the sedimentary rocks that Curiosity studied don't look any different from igneous rocks. This implies that the original source region for the sediment did not spend a lot of time exposed to water alteration before the sediment was transported into the basin. Chemical analyses of Martian soils also indicate that they contain about 2% by weight of water.

Additionally, images of a blue-black rock, with whitish tones in patterns reminiscent of feldspar crystals, appears similar to terrestrial granitic rocks. The Curiosity team looked specifically at areas scrubbed free of Mars's ubiquitous dust, including the flanks of Mars's only dust-free volcano, in Syrtis Major, and suggest that similar rocks are also to be found here. This implies that Martian igneous rocks may have differentiated further than had previously been thought.

Currently Curiosity is making tracks towards Mount Sharp, in Gale crater. In the distant past sedimentary layers were deposited in this vicinity, during a stage when the Red Planet was far wetter and warmer than today, and thus more hospitable to the origin of life. The huge mountain rises about 5.5 km from the centre of Gale Crater. The overland journey could take nearly a year or even longer into 2014 to arrive the base of the mountain.

Mars, therefore, is going to be a busy place in 2014, for two new probes are expected to make it into Martian orbit in September, adding more data to that from NASA's Curiosity and Opportunity rovers. Both India's Mars orbiter Mangalyaan and NASA's MAVEN mission are expected to get into their orbits around Mars in September of 2014. MAVEN is going to investigate the Martian atmosphere and hopefully help piece together the history of how the Red Planet lost its atmosphere. Mangalyaan, India's first Mars probe, is designed to beam back images of Mars's surface and hunt for methane in the planet's atmosphere.

The Mars Orbiter Mission (MOM) or Mangalyaan is designed to study Mars's surface features, morphology, mineralogy, and atmosphere. The spacecraft is equipped with five scientific instruments (camera, thermal IR imaging spectrometer, methane sensor, exospheric neutral composition analyzer, and Lyman-alpha photometer) and will nominally spend 6–10 months orbiting and making measurements at Mars.

NASA's MAVEN Mars probe was launched successfully in November 2013. This probe is designed to study the Martian atmosphere while orbiting Mars with a view to determining how the Martian atmosphere and water, presumed to have once been substantial, were lost over time.

China became the third country to make a soft landing on the Moon with their Chang'e 3 lander at the end of 2013.Despite initial problems with communication it is hoped that its science programme will develop during 2014. The Yutu rover deployed by the lander in Mare Imbrium will use a robotic arm to collect lunar dust samples for analysis, and it will beam back photos of the Moon's surface to Earth. Images already sent back show the lander on the lunar surface (Figure 13.1).

The Visible/Infrared Spectrometer made its first data acquisition session on 23 December 2013, followed by another on the 24th for a total of 54 minutes of operation that provided

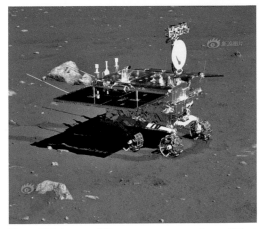

Figure 13.1 Image of Chang-e rover on surface of Mare Imbrium. Courtesy Chinese Academy of Sciences.

good data. On 25 December, Yutu used its APXS to acquire the first X-ray fluorescence spectrum of the lunar regolith. The spectrum released by the Institute of High-Energy Physics shows prominent peaks of Mg, Al, Si, K, Ca, Ti, Cr and Fe which are known to be abundant on the Moon, forming common lunar minerals such as plagioclase feldspar, pyroxene, olivine, ilmenite and others. Also present in the spectrum are indications of the minor elements Sr, Y and Zr, which have also been detected in samples brought back from the Moon by Apollo.

New Horizons is a mission designed to fly by Pluto and its moon Charon and transmit images and data back to Earth and is due to be launched in early July 2014. It will then continue on into the Kuiper Belt. The primary objectives are to characterize the global geology and morphology and map the surface composition of Pluto and Charon, and characterize the neutral atmosphere of Pluto and its escape rate. Other objectives include studying the time variability of Pluto's surface and atmosphere, imaging Pluto and Charon in stereo, mapping the terminators and composition of selected areas of Pluto and Charon at high-resolution, characterizing Pluto's upper atmosphere, ionosphere, energetic particle environment, and solar wind interaction. It is also intended that the probe will search for an atmosphere around Charon.

In August 2014, the European Space Agency's Rosetta Spacecraft will rendezvous with Comet 67P/Churyumov-Gerasimenko and deploy its Philae lander. Rosetta is designed to rendezvous with Comet 67 P/Churyumov-Gerasimenko, drop a probe on the surface, study the comet from orbit, and fly by at least one asteroid en route. The principal goals are to study the origin of comets, the relationship between cometary and interstellar material and its implications with regard to the origin of the Solar System.

The Dawn spacecraft is due to be launched in early 2015 and is a mission designed to rendezvous and orbit the asteroids 4 Vesta and 1 Ceres. The scientific objectives of the mission are to characterize the asteroids' internal structure, density, shape, size, composition and mass and to return data on surface morphology, cratering, and magnetism. These measurements will help determine the thermal history, size of the core, role of water in asteroid evolution and what meteorites found on Earth come from these bodies, with the ultimate goal of understanding the conditions and processes present at the solar system's earliest epoch and the role of water content and size in planetary evolution. The data returned will include, for both asteroids,

full surface imagery, full surface spectrometric mapping, elemental abundances, topographic profiles, gravity fields, and mapping of remnant magnetism, if any.

Juno is a NASA New Frontiers mission to the planet Jupiter. It was launched from Cape Canaveral in 2011 and will arrive in July 2016. The spacecraft is to be placed in a polar orbit to study Jupiter's composition, gravity field, magnetic field, and polar magnetosphere. Juno will also search for clues about how it formed, including whether Jupiter has a rocky core, the amount of water present within the deep atmosphere, and how its mass is distributed. It will also study its deep winds. As of 14 February 2014, Juno was approximately 215 million kilometres from Earth, travelling at a velocity of about 24 kps relative to the sun. The spacecraft will orbit Jupiter 33 times during one Earth year.

Finally, NASA is planning to launch a test flight of their *Orion* spacecraft — a new crew-carrying capsule expected to bring astronauts to Mars and other deep space destinations — in September 2014. For the test flight, Orion will plunge through the atmosphere so that engineers can assess the efficiency of the capsule's heat shield. Orion is also expected to make a parachute-aided splashdown in the Pacific Ocean. This unpiloted test flight will be launched atop a Delta IV Heavy rocket from Florida's Cape Canaveral.

The next few years promise much new and interesting data from a variety of missions. Let us hope that success meets all of them and that planetary science can take a few more forward steps in our understanding of the amazing Solar System.

Appendix 1

Phobos and Deimos

The planet Mars has two small moons. Phobos has a longest dimension of 27 km and a mean density of 1.9 g cm^{-3}, and Deimos, which measures 11 × 15 km, has a density just over half that of Phobos. The surface of Phobos is pockmarked with impact craters and is evidently severely brecciated (Figure Ap.1). The largest crater, Stickney, is 10 km

radiating grooves associated with it. Large blocks have been imaged on its rim, the largest of which is about 50 metres across. Shallow grooves are also visible on high resolution images (Figure Ap.2).

Figure Ap.2 Mars Global Surveyor image of a part of the rim of Stickney, showing boulders on its slopes. The largest of these is 50 m across. PIA01336. Courtesy NASA/JPL/Malin Space Science Systems.

Figure Ap.1 Voyager image of Phobos, showing the impact crater, Stickney, and its radiating grooves. VO1-08. Courtesy NASA/JPL.

across and its impact must have come close to destroying the moon. It covers roughly one-fifth of Phobos's area and has a series of

Deimos is rather different, in general being less heavily cratered than its larger companion. It has a low albedo and appears to be mantled in non-reflective dust of some kind. Some images show bright bands radiating away from impact depressions (Figure Ap.3). Its density (1.8 g cm⁻3) is rather low for it to be a chondritic object, as is the case with the somewhat denser Phobos, but both moons may have relatively high porosity, which could account for this. Both also were clearly captured by Mars early on in its history.

Figure Ap.3 Voyager 2 image of the moon, Deimos, showing small craters and a somewhat smoother surface than that of Phobos. VO2-09. Courtesy NASA/JPL.

Appendix 2

Planetary stratigraphic timescales

The stratigraphic timescale that pertains to the Earth (Palaeozoic, Mesozoic, Cenozoic, etc.) is based on study of rock successions, the fossils they contain and radiometric dating. It is well constrained. That erected for the Moon (Nectarian, Imbrium, Erastosthenian, etc.) has its basis in the superposition of ejecta associated with the larger impact basins and has been constrained by a number of radiometric dates for returned samples. The nomenclature proposed for Mars (Noachian, Hesperian, Amazonian, etc.) has its basis in the impact crater counts made for different surfaces that can be recognized in high resolution imagery. No absolute ages are available for this planet. The reader is advised to refer to the further reading list if further clarification is required for individual bodies.

Glossary

A

Accretion [7]: accretion refers to the collision and adhesion of cooled microscopic dust and ice particles electrostatically, eventually leading to the formation of planetesimals.

accretionary prism [56]: wedge-shaped mass of sedimentary material accumulated and deformed in a subduction zone close to a continental margin.

adiabatic gradient [7]: the adiabatic gradient is the rate of increasing temperature with respect to increasing depth in the Earth's interior. Away from tectonic plate boundaries, it is about 25°C per kilometre of depth.

allotropes [106]: the term allotrope refers to one or more forms of an elementary substance.

anorthosite [49]: a type of plutonic igneous rock predominantly composed of plagioclase feldspar. It is the dominant material of the Moon's highland crust.

aphelion [11]: the point in its elliptical orbit around the Sun where it is at its greatest distance.

arachnoid [75]: generally circular volcanic structure defined by annular fractures, found only on Venus.

asteroid [1]: small rocky bodies, most of which orbit between the orbits of Mars and Jupiter.

asthenosphere [22]: a 100–200 km thick semifluid layer of the Earth, below the outer rigid lithosphere, forming part of the mantle and thought to be able to flow vertically and horizontally.

astronomical unit (AU) [3]: a unit of measurement equivalent to the mean distance between Earth and Sun. It is equal to 149 600 000 km.

aurorae [29]: light phenomena produced in the upper atmosphere by charged particles emitted from the Sun.

B

barycentre [10]: the centre of mass about which a system rotates.

basalt [39]: a volcanic rock composed primarily of clinopyroxene, plagioclase feldspar and olivine. It forms the Earth's oceanic crust.

C

C1-carboniferous chondrite [6]: a type of carbonaceous chondrite that is strongly magnetic, has a lower density than the other two types, contains sulphates, and has a carbon content of about 3.5%.

c-type asteroids [9]: a class of carbonaceous rocky bodies. They are the most common variety, forming around 75% of known asteroids. C-type asteroids are extremely dark, with albedos typically in the 0.03 to 0.10 range.

chalcophile [18]: elements which have a strong affinity for S, Se, and Te. They commonly form ores

chondrite [3]: a form of primitive stony meteorite containing spherical grains termed chondrules. They are all extremely old.

chondrules [6]: small, rounded particles embedded in most chondritic stony meteorites. Chondrules are usually about one millimetre in diameter and consist largely of the silicate minerals olivine and pyroxene. From textural and chemical relationships, it is clear that they were formed at high temperatures as dispersed molten droplets that subsequently solidified and aggregated into chondritic masses.

comet [1]: small icy bodies that, when passing close to the Sun, display a visible atmosphere or coma and sometimes also a tail. These phenomena are due to the effects of solar radiation and the solar wind upon the comet's nucleus. Comet nuclei range from a few hundred metres to tens of kilometres across and are composed of loose collections of ice, dust, and small rocky particles.

cometary coma [19]: the cloud of dust and gas surrounding a comet's nucleus.

continental drift [30]: the movement of the Earth's continents relative to one another, a process driven by plate tectonics. The notion that this might occur was first put forward by Abraham Ortelius in 1596, but was investigated more fully by Alfred Wegener in 1912.

Coriolis effect [30]: the deflection of fluid particles on the Earth induced by its axial rotation.

cryovolcanism [00]: volcanic activity involving icy compounds.

cumulate [49]: an igneous rock in which early-formed crystal phases have sunk to concentrate specific chemical elements such as Mg and Fe.

D

diapir [75]: tear-shaped body of buoyant molten magma.

differentiation [17]: a process whereby the different constituents of a body are separated out into compositionally distinct layers. The term also applies to fractional crystallization of a melt, whereby the removal and segregation of mineral precipitates changes the composition of the melt.

dipole [27]: as used in geology, a dipole refers to a magnetic dipole. A permanent magnet, such as a bar magnet, owes its magnetism to the intrinsic magnetic dipole moment of the electron. The two ends of a bar magnet are referred to as poles, and are labelled 'north' and 'south'. The dipole moment of the bar magnet points from its magnetic south to its magnetic north pole. The north pole of a bar magnet in a compass points north. However, this means that Earth's geomagnetic north pole is the south pole of its dipole moment, and vice versa.

dwarf planet [1]: the International Astronomical Union defines a dwarf planet as a celestial body orbiting a star that is massive enough to be rounded by its own gravity but has not cleared its neighbouring region of planetesimals. More explicitly, it has to have sufficient mass to overcome its compressive strength and achieve hydrostatic equilibrium.

E

eccentricity [10]: the eccentricity (e) of a body is the amount by which its path around another object departs from circularity. A circular orbit has $e = 0$, while an elliptical one has $e = <1$. Parabolic orbits have $e = 1$, while hyperbolic orbits have $e = >1$.

ecliptic [1]: a circle projected onto the celestial sphere, representing the Sun's annual path relative to the background stars. It actually represents the intersection of the Earth's orbital plane with the celestial sphere and, because the Earth is inclined at 23.5° to the orbital plane, the ecliptic is inclined to the celestial equator by the same amount.

electromagnetic radiation [2]: electro-magnetic radiation (EMR) is a form of energy emitted and absorbed by charged particles which exhibits wave-like behaviour as it travels through space. EMR has both electric and magnetic field components. In a vacuum, EMR propagates at a characteristic speed, the speed of light.

exosphere [42]: the outermost region of a planet's atmosphere.

F

fire fountaining [87]: fluid eruptions often send up tall fountains of molten lava that fall back in and around the eruptive vent. This is termed fire fountaining.

flood lavas [69]: flood lavas emerge from long systems of fissures, as opposed to centralized vents. Generally they are basaltic in character and cover large areas of planetary surfaces. On Earth they are represented in the Columbia River and Deccan regions.

fretted channels [63]: the complex mix of cliffs, mesas and buttes and straight-walled and sinuous flat-floored canyons that straddles the dichotomy boundary of Mars. The debris flows on their floors is a manifestation of mass movements.

frost line [92]: the distance from the Sun beyond which icy compounds could remain solid within the solar nebula.

fusion reactions [2]: Nuclear fusion is a reaction in which two or more atomic nuclei collide at a very high speed and join to form a new type of atomic nucleus. During this process, matter is not conserved because some of the mass of the fusing nuclei is converted to photons (energy). Fusion is the process that powers active or 'main sequence' stars.

G

galaxy [1]: a large mass of stars forming systems that may contain between 10^6 and 10^{12} stars. The star system of which the Sun is a member is known as the Milky Way Galaxy.

gravitational potential energy [20]: the energy stored in an object as the result of its vertical position or height. The energy is stored as the result of the gravitational attraction of another body for the object. There is a direct relation between gravitational potential energy and the mass of an object. More massive objects have greater gravitational potential energy.

H

heliopause [27]: the distance from the Sun where the outgoing solar wind is exactly balanced by the interstellar medium.

heliosphere [27]: the heliosphere is a region of space dominated by the Sun, a sort of bubble of charged particles in the space surrounding the Solar System. These particles are expelled from the sun by the solar wind and into the interstellar medium.

hot spot [58]: long-lived volcanic foci are believed to sit above hot spots. Originally thought to be caused by a narrow stream of hot mantle convecting up from the mantle–core boundary, the latest evidence points to upper-mantle convection as a cause.

hydrogen deuteride [98]: a diatomic molecule composed of the two isotopes of hydrogen: the majority isotope 1H protium and 2H deuterium. Its molecular formula is HD. It is one of the minor but noticeable components of the atmospheres of all the giant planets.

I

impact basin [42]: a large impact structure with a diameter in excess of 300 km. There are over 40 on the Moon. They often have a concentric ring structure and an extensive ejecta blanket.

interplanetary magnetic field [27]: the magnetic particles that emanate from the Sun generate a strong magnetic field that extends throughout the Solar System.

ionosphere [41]: the ionosphere represents less than 0.1% of the total mass of the Earth's atmosphere; however, it is extremely important as it makes long-distance radio transmissions possible. All planets have such a layer at the very edge of their atmosphere; it is ionized by solar radiation. Aurorae have their origin in this layer.

K

Kepler mission [3]: this mission is a space observatory launched by NASA to discover Earth-like planets orbiting other stars. The spacecraft was launched on March 7, 2009.

komatiite [25]: Komatiites are unusual Mg-rich igneous rocks erupted in the Earth's early history. The melting temperatures of komatiite magmas are exceedingly high (~1700 oC) but become lowered by the addition of water.

Kuiper Belt [2]: a region of space with an elliptical form. In many ways it is similar to the asteroid belt, but the Kuiper Belt is composed almost entirely of icy materials. It is located approximately 30 to 50 AU (astronomical units) from the Sun.

L

Late Heavy Bombardment [5]: Lunar impact cratering data suggests that a cataclysmic spike in the cratering rate occurred 4.1 to 3.8 Gya, after the planets formed; this event is known as the Late Heavy Bombardment. There is much discussion on the validity of this thesis.

Laws of Planetary Motion [13]: three laws were devised by Johannes Kepler to define the mechanics of planetary motion. The first law states that planets move in elliptical orbits, with the Sun being one focus of the ellipse. The second law states that the radius of the vector joining the planet to the Sun sweeps out equal areas in equal times as the planet travels around the ellipse. The third law states that the ratio of the squares of the orbital period for two planets is equal to the ratio of the cubes of their mean orbit radius.

libration [12]: the apparent tilting of the Moon as seen from the Earth allows observers to observe 59% of the lunar surface (although only 50% at any one time).

lithophile [18]: lithophile elements are those which combine with oxygen and therefore reside near to inner planetary surfaces. One group includes elements having large ionic radius, such as potassium, rubidium, caesium, strontium and barium. These are called LILE, or large-ion lithophile elements.

lithosphere [55]: the outer region of the earth, consisting of the crust and upper mantle, approximately 100 km thick. The lithosphere remains rigid for very long periods of time, in which it deforms elastically and through brittle failure. The lithosphere is broken into tectonic plates.

Lunar maria [69]: the extensive basaltic floods that partly filled the larger impact basins and craters on both sides of the Moon.

M

m-type asteroids [9]: m-type asteroids are moderately bright and are usually metal but sometimes metal-stone mixtures. Some of them have a very similar composition to iron meteorites that have fallen to Earth. They are believed to have come from the cores of differentiated planetoids that were later broken apart.

magma ocean [20]: the huge amount of accretional energy imparted to the inner planets and Earth's Moon during the early stages of Solar System evolution liberated sufficient heat for a large portion of the bodies to become completely molten, forming a magma ocean.

magnetic excursion [30]: a failed attempt at a complete magnetic reversal.

magnetic inclination [31]: the angle at which a freely-suspended magnetic needle will hang when suspended in the Earth's magnetic field. The angle is related to the latitude at which the needle lies.

magnetic reversal [30]: a period in the geological record during which the ambient field flips to the opposite polarity. Sometimes the field almost reverses but then reverts; this is known as a magnetic excursion.

magnetopause [34]: this is the abrupt boundary between the magnetosphere and the external currents, such as the solar wind.

magnetosphere [27]: a magnetosphere is the area of space near a body in which charged particles are controlled by that body's magnetic field. Near the surface of the object, the magnetic field lines resemble those of an ideal magnetic dipole. Farther away from the surface, the field lines are significantly distorted by external currents.

main-sequence star [3]: a star that lies on the main evolutionary band when spectral class is plotted against luminosity on what is known as the Hertsprung-Russell diagram.

mantle plume [58]: an upwelling of abnormally hot rock within the Earth's mantle. As the heads of mantle plumes can partially melt when they reach shallow depths, they are thought to be the cause of volcanic centres known as hotspots.

mascon [90]: concentrations of mass found on the Moon beneath basalt-filled impact basins.

mesosphere [22, 41]: the portion of the Earth's atmosphere from about 30 to 80 km above the surface, characterized by temperatures that decrease from 10°C to -90°C with increasing altitude. Similar layers occur on the other planetary bodies.

meteorite [2]: when a meteoroid passes

through the Earth's atmosphere, it is known as a meteorite. It may survive and impact the surface or burn up in the atmosphere.

meteoroid [1]: a solid fragment of planetary material from space that probably derived from an asteroid or comet.

molecular cloud [4]: this is a large region of cool molecular hydrogen, often located in the arms of spiral galaxies, where stars are believed to be born.

N

nebula [4]: cloud of gas and dust in interstellar space.

novae [75]: circular volcanic structures that are unique to Venus.

O

Oort Cloud [2]: the Oort Cloud is composed of a vast number of icy objects that are predicted to envelop the Solar System far beyond the orbit of Neptune and the Kuiper Belt. The cloud has been estimated to extend out to a distance of 50 000 AU from the Sun and marks the edge of the Solar System.

orbital resonance [5]: an orbital resonance occurs when two orbiting bodies exert a regular, periodic gravitational influence on each other, usually due to their orbital periods being related by a ratio of two small integers. Orbital resonances greatly enhance the mutual gravitational influence of the bodies.

outflow channels [62]: extensive channel networks found on Mars that were cut early in Mars's history by massive floods.

P

paleomagnetism [29]: the study of the magnetism fossilized in terrestrial rocks. It has enabled geologists to establish the position of continents in times past.

perihelion [11]: the point of closest approach of a planetary body to the Sun.

photosphere [6]: the bright illuminated surface of the Sun.

planet [1]: an astronomical object that orbits a star or stellar remnant.

planetesimals [9]: the term applies to fragments of dust and ice that collect in protoplanetary disks such as that within the Solar Nebula.

plate tectonics [54]: the modern theory that explains the geological processes of Earth, driven by convective motions in the mantle layer.

precession [12]: the slow periodic changes in the orientation of a planet's axis of rotation due to the gravitational attraction of the Sun and other bodies.

pressure release melting [56]: a process in which rocks very close to their melting point melt, due to the release of overlying pressure.

pulsar [3]: a pulsar is a highly magnetized, rotating neutron star that emits a beam of electromagnetic radiation.

pyrolite [68]: a theoretical rock considered to very closely approach the composition of the Earth's mantle.

R

radiometric dating [57]: the technique of calculating a rock's age by the amount of decay experienced by radioactive elements trapped in the lattices of silicate minerals.

radionuclides [20]: a radionuclide, or radioactive nuclide, is an atom with an unstable nucleus, characterized by excess energy available to be imparted to a newly created particle. A radionuclide – also referred to as a radioisotope – is said to undergo radioactive decay. Those with suitable half-lives play an important part in dating rocks.

refractory compounds [30]: chemical compounds that are capable of resisting or withstanding very high temperatures.

regolith [87]: the loose surface layer of a planetary surface. This differs from soil, which generally implies organic content, as on Earth.

retrograde motion [11: the motion of a body from east to west, contrary to the prevailing direction of movement within the solar system.

ridge belts [53]: ridge belts are composed of closely spaced individual ridges 5–20 km wide that form sinuous landforms within the northern plains of Venus. They may be up to 400 km wide and 2000 km in length.

ring system [2]: several of the outer planets have ring systems. Consisting of countless small particles of ice and rock, these range in size from micrometres to metres. The rings of Saturn are the most extensive, and within them are several gaps swept clean by small moons that are embedded in the rings.

S

s-type asteroids [9]: this kind of asteroid accounts for about 17% of known asteroids. They are relatively bright and have a composition of metallic iron mixed with iron- and magnesium-silicates. They dominate the inner asteroid belt.

sea-floor spreading [57]: the process whereby new oceanic crust forms at mid-oceanic ridges. First confirmed by paleomagnetic studies, this is now a fundamental part of modern plate tectonics theory.

sidereal day [18]: the sidereal day is equivalent to the rotation period of the celestial sphere.

siderophile [18]: siderophile elements are the high-density transition metals which tend to sink into the core because they dissolve readily in iron either as solid solutions or in the molten state. They have virtually no affinity for oxygen.

Solar nebula [4]: the Solar System began forming within a concentration of interstellar dust and hydrogen gas called a molecular cloud. The cloud contracted under its own gravity and our proto-Sun formed in the hot, dense centre. The remainder of the cloud formed a swirling disk called the solar nebula.

Solar System [1]: the Sun and its family of planets and other minor bodies, including comets, asteroids, and meteorites.

Solar wind [5]: the stream of charged particles that emanates from the Sun.

star [1]: a self-luminous sphere of hot gas. Most shine because of the fusion of elements in their core regions.

stratosphere [41]: the second major layer of a planet's atmosphere, just above the troposphere, and below the mesosphere. It is stratified in temperature, with warmer layers higher up and cooler layers farther down.

subduction trench [56]: deep trenches are generated as oceanic crust is subducted beneath a continental margin. Such trenches may be filled with thick sequences of deformed marine sediments.

subduction zone [56]: a zone on the Earth where two tectonic plates converge. The denser oceanic plate descends below the continental margin as an inclined plane – the subduction zone. This phenomenon generates intense seismic activity.

supernova [4]: a hugely energetic stellar explosion.

synchronous rotation [10]: the situation where a planet's orbital period is identical with that of its rotation.

T

terrestrial planets [2]: the inner worlds of the Solar System; relatively small and dense and composed of silicate rocks and metallic iron.

tesserae [53]: belts of deformed rocks that show strong evidence for parallel folding and faulting and may form primarily by compression, deformation, and uplift of the lithosphere of Venus.

thermokarst [60]: a landscape characterized by very irregular surfaces of marshy hollows and small hummocks formed as ice-rich permafrost thaws. Typically this occurs on Earth in Arctic areas, and on a smaller scale in mountainous areas.

thermosphere [44]: the outermost of a planet's atmospheric layers.

tholins [113]: molecules formed by solar ultraviolet irradiation of simple organic compounds such as methane or ethane. Tholins do not form naturally on modern-day Earth, but are found in great abundance on the surface of icy bodies in the outer Solar System.

tropopause [41]: the layer in the terrestrial atmosphere (or other planetary atmosphere) between the troposphere and the stratosphere. It is the point where air ceases to cool with height, and becomes almost completely dry.

troposphere [41]: the lowest layer of a planet's atmosphere. On Earth this extends upwards to between 7 and 20 km and accounts for between 75 and 80% of the total mass.

T Tauri star [5]: T Tauri is the prototype for a class of very young stars, still in the process of gravitational contraction and not yet on the Main Sequence.

U

Universe [1]: a definition for all matter – stars, galaxies, planets and intergalactic space.

V

valley networks [63]: valley networks are branching networks of valleys on Mars that superficially resemble terrestrial river drainage basins. They are located mainly in the terrain of the Martian southern highlands and are geologically very ancient.

Van Allen radiation belts [29]: zones of charged particles that surround the Earth. The inner zone is composed mainly of protons, and the outer of electrons.

volcanic rises [53]: on Venus, major volcanic rises are believed to have formed above long-lived hotspots in the mantle layer. These complex elevated areas represent regions of tectonic and volcanic activity.

W

wrinkle ridges [60]: these are landforms found on lunar maria. They are low, sinuous ridges formed on the mare surface that can extend for up to several hundred kilometres. They are tectonic features created when the basaltic lava first cooled and contracted. Similar features are found on the other rocky planets.

Further reading

Baker, David. *NASA Mars Rovers Manual: 1997-2013* (Sojourner, Spirit, Opportunity and Curiosity) (Owners' Workshop Manual). Haynes, 2013.

Barlow, Nadine. *Mars: An Introduction to its interior, surface and atmosphere.* CUP, 2008.

Bougher, Stephen, W., Hunten, Donald M. and Phillips, Roger J. *Venus II: Geology, Geophysics, Atmosphere and Solar Wind Environment.* University of Arizona Press, 1997.

Byrne, Charles. *The Far side of the Moon. A photographic Guide.* (Patrick Moore Practical Astronomy Series).

Cattermole, Peter. *Planetary Volcanism.* Wiley/Praxis, 2nd edition, 1996.

Cattermole, Peter. *Venus: The geological story.* UCL, 1994.

Chapman, Mary (ed). *The Geology of Mars* CUP, 2011.

Emiliani, Cesare. *Planet-Earth: Cosmology, Geology and the evolution of life and the environment.* CUP, 1992.

Encrenaz, Thérèse, Kallenbach, R., Owen, T. and Sotin, Christophe. *The Outer Planets and their Moons: Comparative Studies of the Outer Planets prior to the Exploration of the Saturn System by Cassini-Huygens* (Space Sciences Series of ISSI). Springer, 2005.

Encrenaz, Thérèse. *Planets: Ours and others – from Earth to exoplanets.* World Scientific, 2013.

Greeley, Ronald. *Introduction to Planetary Geomorphology.* CUP, 2012.

Greeley, Ronald. *Planetary Landscapes.* Chapman and Hall, 2nd edition, 1994.

Harland, David M. *Cassini at Saturn: Huygens Results.* Springer/Praxis Space Exploration, 2007.

Jones, Eric. *On the Moon: The Apollo Journals,* Springer Praxis Books, 2007.

Lopes, Rosaly and Carroll, Michael. *Alien Volcanoes,* John Hopkins, 2008.

Moore, Patrick and Rees, Robin. *Patrick Moore's Data Book of Astronomy,* CUP, 2011.

Mutch, Thomas. *Geology of the Moon – A stratigraphic view.* Princeton, 1970.

Ross-Taylor, Stuart. *Solar System Evolution: A new perspective.* CUP, 2001.

Rothery, David. *Satellites of the Outer Planets: Worlds in their own right,* OUP, 1999.

Shirley, J. H. And Fairbridge, Rhodes, W. (eds). Encyclopaedia of Planetary Sciences. Chapman and Hall, 1997.

Verba, Joan Marie. *Voyager: Exploring the Outer Planets.* FTL Publications, Minnesota. 2013.

Vita-Finzi, Claudio and Fortes, Andrew Dominic. *Planetary Geology: An Introduction,* 2nd edition, 2013, Dunedin.

Wilhelms, Don E. *The Geologic History of the Moon.* USGS Professional Paper 1348, Washington, 1987.

Useful web sites

http://photojournal.jpl.nasa.gov/
http://www.lpi.usra.edu/
http://sci.esa.int/cassini-huygens/
http://earthobservatory.nasa.gov/
http://www.lpl.arizona.edu/
http://www.lpi.usra.edu/lunar/lunar_images/
http://www.lpi.usra.edu/resources/lunar_orbiter/
http://www.mentallandscape.com/C_CatalogMoon.html
http://ser.sese.asu.edu/MOON/clem.html
http://solarsystem.nasa.gov/index.cfm